Standard Grade | General

Mathematics

Leckie×Leckie

First exam published in 2001.

Published by Leckie & Leckie, 8 Whitehill Terrace, St. Andrews, Scotland KY16 8RN tel: 01334 475656 fax: 01334 477392 enquiries@leckieandleckie.co.uk www.leckieandleckie.co.uk

ISBN 1-84372-307-7

A CIP Catalogue record for this book is available from the British Library.

Printed in Scotland by Scotprint.

Leckie & Leckie is a division of Granada Learning Limited, part of ITV plc.

Acknowledgements

Leckie and Leckie is grateful to the copyright holders, as credited at the back of the book, for permission to use their material. Every effort has been made to trace the copyright holders and to obtain their permission to use their copyright material. Leckie & Leckie will gladly receive information enabling them to rectify any error or omission in subsequent editions.

[BLANK PAGE]

G

FOR OFFICIAL USE

	KU	RE
Total marks		

2500/403

NATIONAL
QUALIFICATIONS
2001

WEDNESDAY, 16 MAY
10.40 AM – 11.15 AM

MATHEMATICS
STANDARD GRADE
General Level
Paper 1
Non-calculator

Fill in these boxes and read what is printed below.

Full name of centre

Town

Forename(s)

Surname

Date of birth
Day Month Year

Scottish candidate number

Number of seat

1 **You may not use a calculator.**

2 Answer as many questions as you can.

3 Write your working and answers in the spaces provided. Additional space is provided at the end of this question-answer book for use if required. If you use this space, write clearly the number of the question involved.

4 Full credit will be given only where the solution contains appropriate working.

5 Before leaving the examination room you must give this book to the invigilator. If you do not you may lose all the marks for this paper.

SCOTTISH
QUALIFICATIONS
AUTHORITY

FORMULAE LIST

Circumference of a circle: $C = \pi d$

Area of a circle: $A = \pi r^2$

Curved surface area of a cylinder: $A = 2\pi rh$

Volume of a cylinder: $V = \pi r^2 h$

Volume of a triangular prism: $V = Ah$

Theorem of Pythagoras:

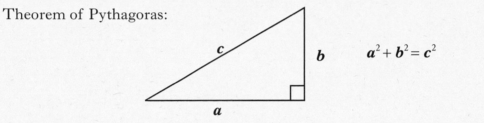

$$a^2 + b^2 = c^2$$

Trigonometric ratios
in a right angled
triangle:

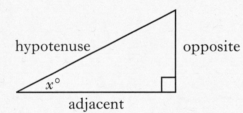

$$\tan x° = \frac{\text{opposite}}{\text{adjacent}}$$

$$\sin x° = \frac{\text{opposite}}{\text{hypotenuse}}$$

$$\cos x° = \frac{\text{adjacent}}{\text{hypotenuse}}$$

Gradient:

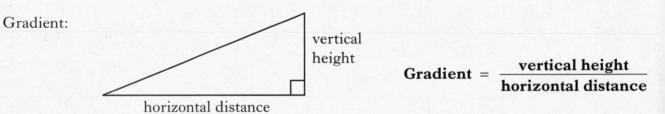

$$\text{Gradient} = \frac{\text{vertical height}}{\text{horizontal distance}}$$

Marks | KU | RE

1. Work out the following.

(a) $18\cdot54 + 0\cdot61 - 5\cdot3$

1

(b) $3\cdot36 \times 70$

1

(c) $0\cdot296 \div 4$

1

(d) $\frac{3}{4}$ of $480\,g$

2

[Turn over

Marks | KU | R

2.

A student pays a train fare of £24.

If this represents 60% of the full adult fare, what is the full adult fare?

3

3. Brian checks the five day weather forecast for Paris.

PARIS – FORECAST for 15 January			
	Maximum (°C)	Minimum (°C)	
Saturday	3	−3	Cloudy
Sunday	2	0	Sunny
Monday	7	4	Cloudy
Tuesday	7	2	Sunny
Wednesday	5	−2	Sunny

Calculate the **mean** minimum temperature for the five day weather forecast.

3

4. (*a*) Write the number $1 \cdot 5 \times 10^{-1}$ in full.

1

(*b*) Mark the position of this number on the number line below.

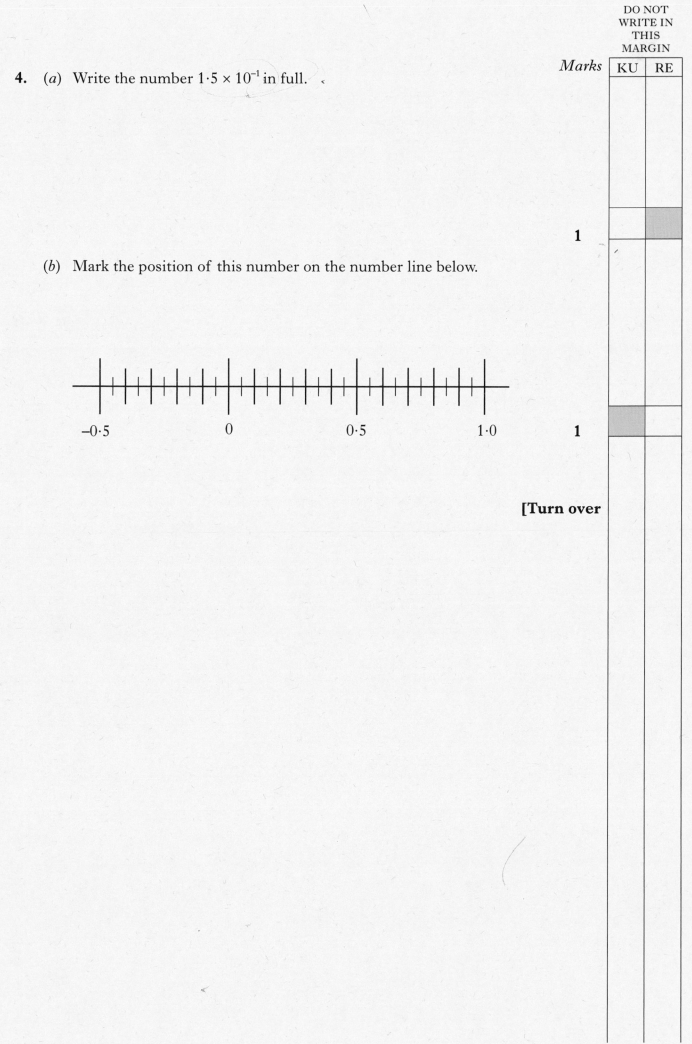

−0·5   0   0·5   1·0

1

[Turn over

Marks | KU | RI

5. A seaside promenade is to be covered with tiles.

All the tiles are shaped like this.

Here is part of the design of tiles.

Draw six more tiles to continue the design.   **3**

6. Two trains run from Aberdeen to London Kings Cross.

They both have **the same journey time.**

	Train 1	**Train 2**
Aberdeen *Depart*	1455	2125
Kings Cross *Arrive*	2229	

Find the arrival time for Train 2 at Kings Cross.

3

[Turn over

7. (*a*) Plot the points A (4,6) and C (4,–2) on this grid.

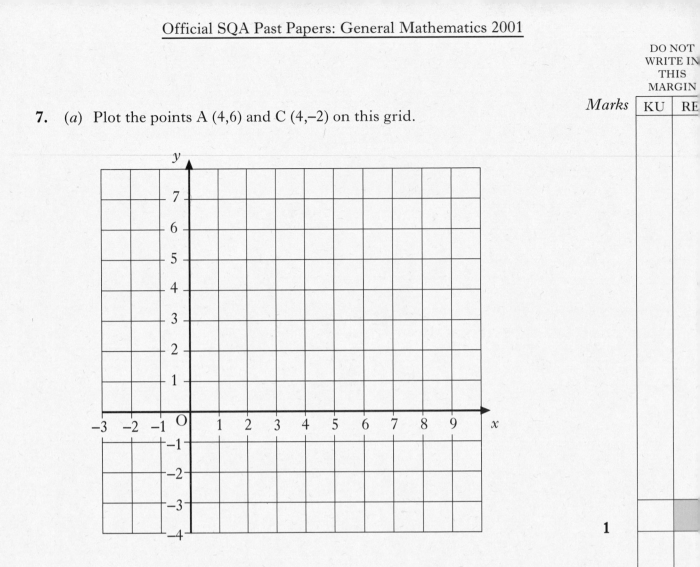

1

(*b*) ABCD is a rhombus with area 24 square units.

Plot B and D on the grid.

3

8.

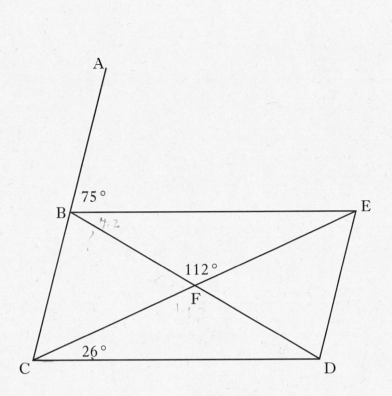

BCDE is a parallelogram.

Angle ABE = 75°, angle ECD = 26°, angle BFE = 112°.

Calculate the size of the angle CBD.

3

[Turn over

Marks

KU | RE

9. There are 1 blue, 2 red and 3 yellow counters in a bag.

(*a*) A counter is taken from the bag.

What is the probability that the counter is red?

1

(*b*) The counter is replaced in the bag and two green counters are added to the bag.

A counter is taken from the bag.

What is the probability that it is **not** yellow?

2

10. At Dunure Tennis and Golf Club, the ratio of tennis players to golfers is 100:350.

(a) Express this ratio in its simplest form.

1

(b) The club has been given £16 200.

This money will be divided between the tennis section and the golf section in the same ratio as above.

How much money will be allocated to the tennis section?

3

[END OF QUESTION PAPER]

ADDITIONAL SPACE FOR ANSWERS

Page twelve

FOR OFFICIAL USE

G

	KU	RE
Total marks		

2500/404

NATIONAL
QUALIFICATIONS
2001

WEDNESDAY, 16 MAY
11.35 AM – 12.30 PM

MATHEMATICS
STANDARD GRADE
General Level
Paper 2

Fill in these boxes and read what is printed below.

Full name of centre

Town

Forename(s)

Surname

Date of birth
Day Month Year

Scottish candidate number

Number of seat

1 **You may use a calculator.**

2 Answer as many questions as you can.

3 Write your working and answers in the spaces provided. Additional space is provided at the end of this question-answer book for use if required. If you use this space, write clearly the number of the question involved.

4 Full credit will be given only where the solution contains appropriate working.

5 Before leaving the examination room you must give this book to the invigilator. If you do not you may lose all the marks for this paper.

SCOTTISH
QUALIFICATIONS
AUTHORITY

FORMULAE LIST

Circumference of a circle: $C = \pi d$
Area of a circle: $A = \pi r^2$
Curved surface area of a cylinder: $A = 2\pi rh$
Volume of a cylinder: $V = \pi r^2 h$
Volume of a triangular prism: $V = Ah$

Theorem of Pythagoras:

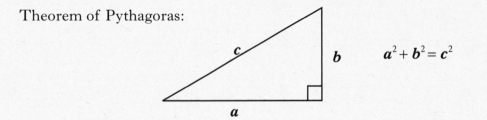

$$a^2 + b^2 = c^2$$

Trigonometric ratios
in a right angled
triangle:

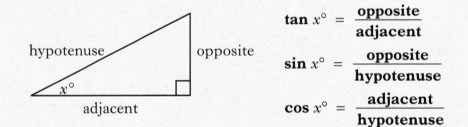

$$\tan x° = \frac{\text{opposite}}{\text{adjacent}}$$

$$\sin x° = \frac{\text{opposite}}{\text{hypotenuse}}$$

$$\cos x° = \frac{\text{adjacent}}{\text{hypotenuse}}$$

Gradient:

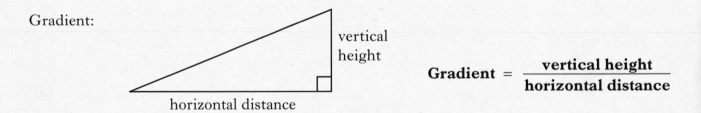

$$\text{Gradient} = \frac{\text{vertical height}}{\text{horizontal distance}}$$

Marks

1. Jayne is 14 years of age and a member of Kelly's Health Club.

   She receives details of next year's subscription rates.

   They are as follows:

Category of member	Payment in full	Payment by instalments
Adult	£390	12 payments of £36
Junior (under 16 years of age)	£195	12 payments of £18
Husband and Wife	£695	12 payments of £65

(a) Jayne decides to pay by instalments.

   How much extra will she pay?

   2

(b) Express this extra cost as a percentage of the payment in full.

   Give your answer correct to 1 decimal place.

   3

2. The number of passengers travelling by bus from Glasgow to Edinburgh was recorded for 20 journeys.

29	45	36	27	41	38	14
48	31	39	24	17	23	34
29	38	42	12	32	36	

(a) Display the information in an ordered stem and leaf diagram.

3

(b) Find the median number of passengers.

1

DO NOT WRITE IN THIS MARGIN

Marks KU RE

3. Shona is planning to buy a new mobile phone.

She knows that she makes between 60 and 120 minutes of calls each month.

Her local phone shop advises that the "All-Talk" or "Talk-Time" tariff are best for her.

They give her the graph below to help her decide.

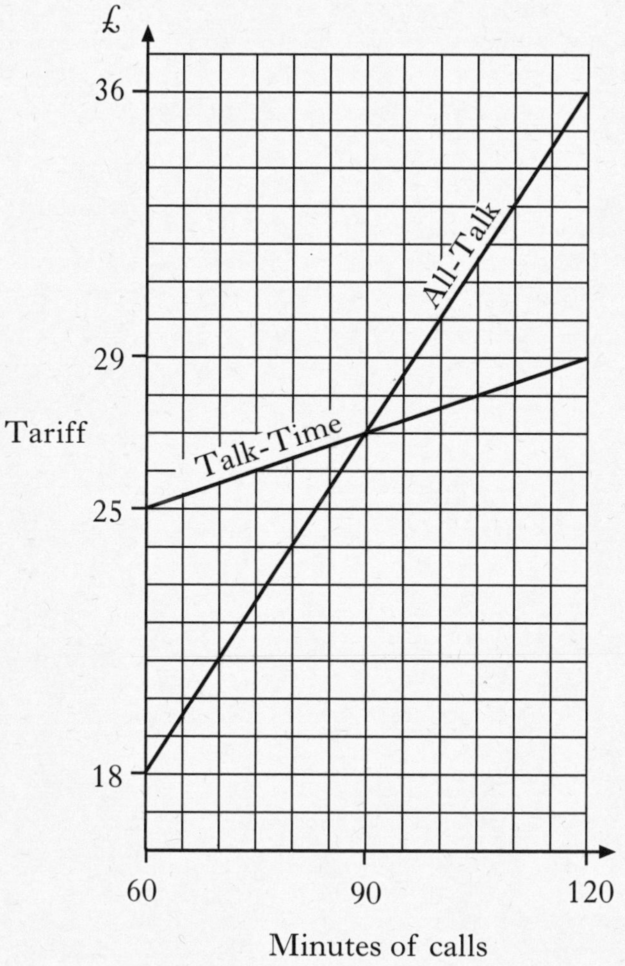

Shona chooses the All-Talk tariff.

Comment on her choice. **2**

[Turn over

Marks | KU | RE

4. A manufacturer has changed its washing powder so that less powder will be needed for each wash.

As a result the new 1·5 kilogram box gives the same number of washes as the old 2 kilogram box.

A family wash used 96 grams of powder from the old 2 kilogram box.

How much powder will be used for a family wash now?

4

Marks | KU | RE

5. Davina sees this advertisment for CAR HIRE while on holiday in Spain.

UNLIMITED MILEAGE, INSURANCE INCLUDED	
Locus Speedster	3100 pesetas per day
	20 000 pesetas per week
A-Drive Trekcar	5560 pesetas per day
	35 000 pesetas per week
ADD 15% TAX	

She decides to hire the Trekcar for 4 days.

Find the cost, in pounds sterling, of hiring the car if the exchange rate is £1= 256 pesetas.

5

[Turn over

Marks | KU | RE

6. Mairi is planning to paint the walls of her room with emulsion paint.

The room is in the shape of a cuboid, with the dimensions shown.

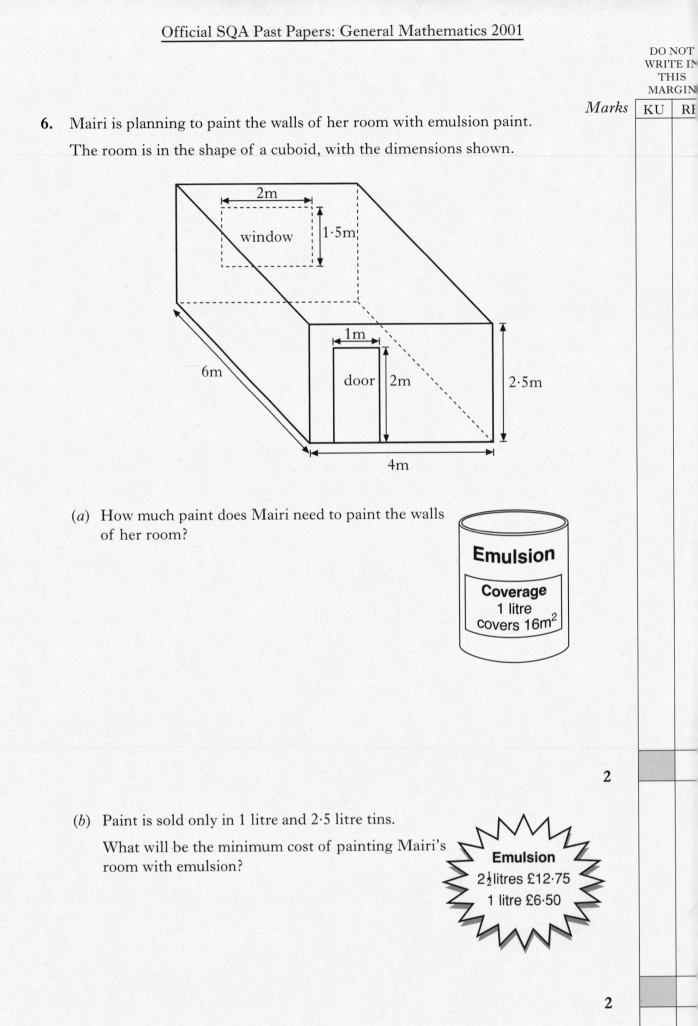

(a) How much paint does Mairi need to paint the walls of her room?

2

(b) Paint is sold only in 1 litre and 2·5 litre tins.

What will be the minimum cost of painting Mairi's room with emulsion?

2

Official SQA Past Papers: General Mathematics 2001

DO NOT
WRITE IN
THIS
MARGIN

Marks | KU | RE

7. John is laying a concrete floor for his garage.

The floor is to be a rectangle 5·5 metres by 3 metres.

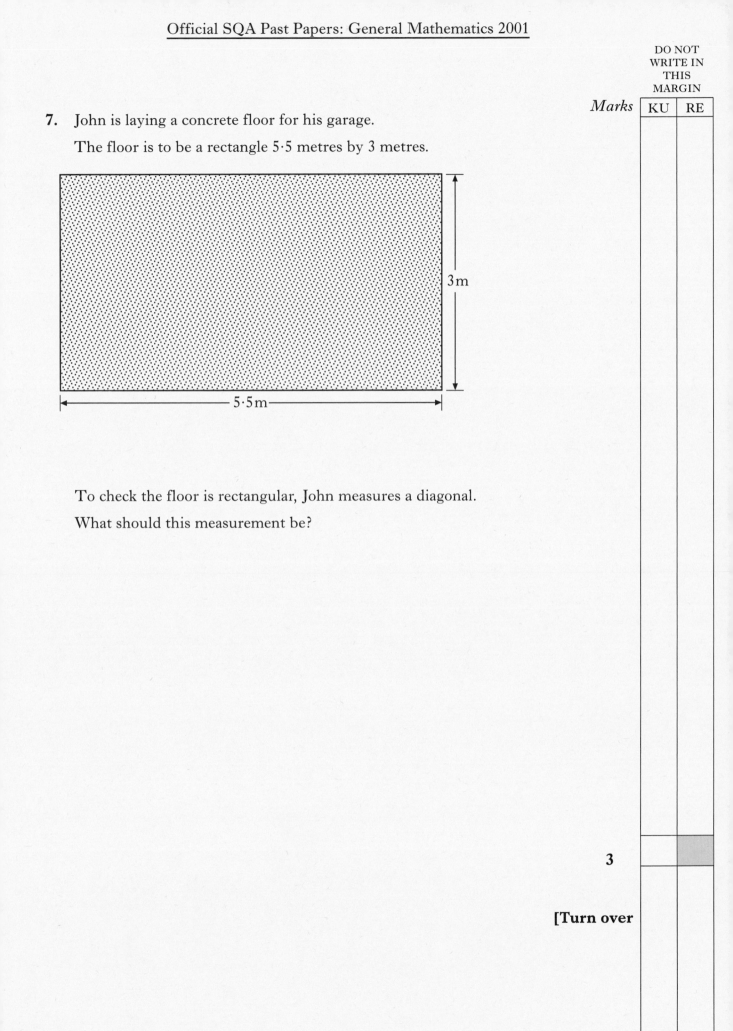

To check the floor is rectangular, John measures a diagonal.

What should this measurement be?

3

[Turn over

8. At the Ewington Athletic Club the length of one lap of the track is 400 metres.

In the 10 000 metres race a runner takes an average of 65·2 seconds to complete each lap.

At this pace, will this runner break the race record of 27 minutes 12 seconds?

4

Marks | KU | RE

9. (*a*) Simplify

$$3(2x + 4) + 4(x - 2).$$

3

(*b*) Solve algebraically the inequality

$$6x + 2 \leq 20.$$

2

[Turn over

10. The base of a round cake tin has the same area as the base of a square cake tin.

The round cake tin has a radius of 10 centimetres.

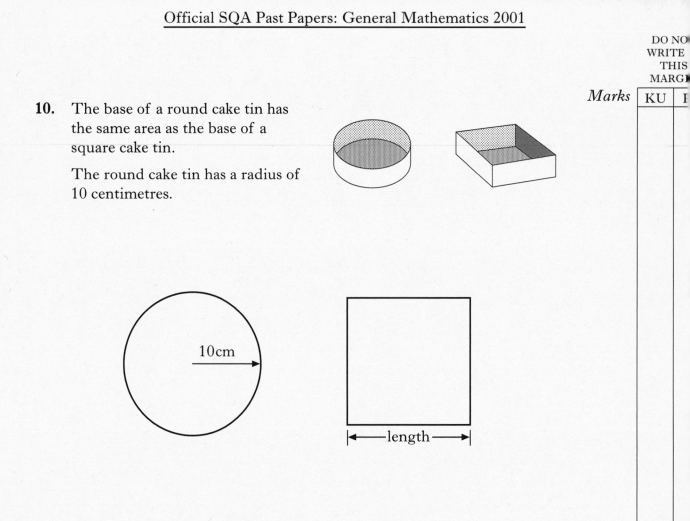

What is the length of the base of the square cake tin?

3

11. (a) The base of a lift is in the shape of a rectangle with a semi-circular end as shown.

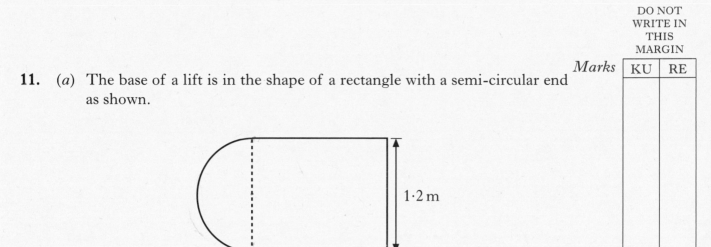

1·2 m

1·4 m

Calculate the area of the base of the lift.

3

(b) The lift is in the shape of a prism and is 220 centimetres high.

Find the volume of the lift.

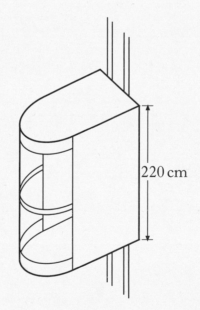

220 cm

2

[Turn over for Question 12 on Page fourteen

12. An architect is designing a room in an attic of a house.

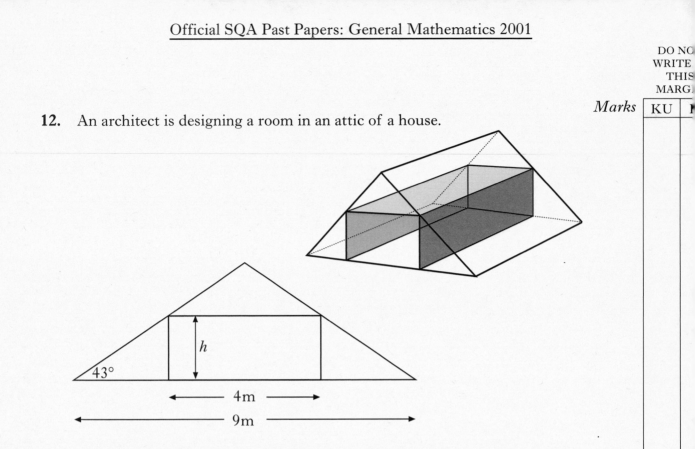

- The room is 4 metres wide.
- The width of the roof is 9 metres.
- The sloping part of the roof makes an angle of 43° with the attic floor.

To satisfy building regulations the height, h, of the room must be **not less than** 2·3 metres.

Does the architect's design satisfy the building regulations?

Give a reason for your answer.

4

[END OF QUESTION PAPER]

[BLANK PAGE]

FOR OFFICIAL USE

G

	KU	RE
Total marks		

2500/403

NATIONAL
QUALIFICATIONS
2002

THURSDAY, 9 MAY
10.40 AM – 11.15 AM

MATHEMATICS
STANDARD GRADE
General Level
Paper 1
Non-calculator

Fill in these boxes and read what is printed below.

Full name of centre

Town

Forename(s)

Surname

Date of birth
Day Month Year   Scottish candidate number   Number of seat

1 **You may not use a calculator.**

2 Answer as many questions as you can.

3 Write your working and answers in the spaces provided. Additional space is provided at the end of this question-answer book for use if required. If you use this space, write clearly the number of the question involved.

4 Full credit will be given only where the solution contains appropriate working.

5 Before leaving the examination room you must give this book to the invigilator. If you do not you may lose all the marks for this paper.

SCOTTISH
QUALIFICATIONS
AUTHORITY

©

FORMULAE LIST

Circumference of a circle: $C = \pi d$

Area of a circle: $A = \pi r^2$

Curved surface area of a cylinder: $A = 2\pi rh$

Volume of a cylinder: $V = \pi r^2 h$

Volume of a triangular prism: $V = Ah$

Theorem of Pythagoras:

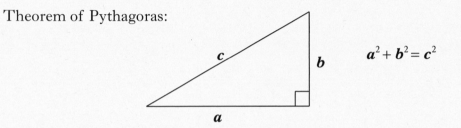

$$a^2 + b^2 = c^2$$

Trigonometric ratios
in a right angled
triangle:

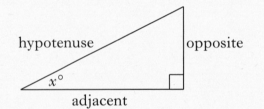

$$\tan x^\circ = \frac{\text{opposite}}{\text{adjacent}}$$

$$\sin x^\circ = \frac{\text{opposite}}{\text{hypotenuse}}$$

$$\cos x^\circ = \frac{\text{adjacent}}{\text{hypotenuse}}$$

Gradient:

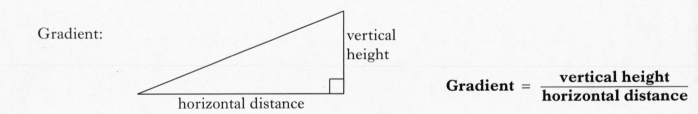

$$\text{Gradient} = \frac{\text{vertical height}}{\text{horizontal distance}}$$

DO NOT
WRITE IN
THIS
MARGIN

Marks | KU | RE

1. Carry out the following calculations.

(a) $9 \cdot 2 - 3 \cdot 71 + 6 \cdot 47$

1

(b) $7 \cdot 29 \times 8$

1

(c) $687 \div 300$

1

(d) $3 \times 2\frac{3}{4}$

2

[Turn over

Marks

2. Davina has a bag of sweets.

It contains three yellow sweets, four purple sweets, two red sweets and six pink sweets.

The corner of her bag is torn and a sweet falls out.

(a) What is the probability that this sweet is yellow?

1

(b) The sweet that fell out was yellow and she put it in a bin.

What is the probability that the next sweet to fall out is pink?

2

3. Complete this shape so that it has quarter-turn symmetry about O.

3

4. There are five million people in the United Kingdom aged 15–19.

30% of these five million people regularly watch cartoons.

How many people is this?

2

[**Turn over**

DO NOT
WRITE IN
THIS
MARGIN

Marks | KU | RE

5. (*a*) On the grid below, plot the points A(−4, −3), B(3, −1) and C(4, 4).

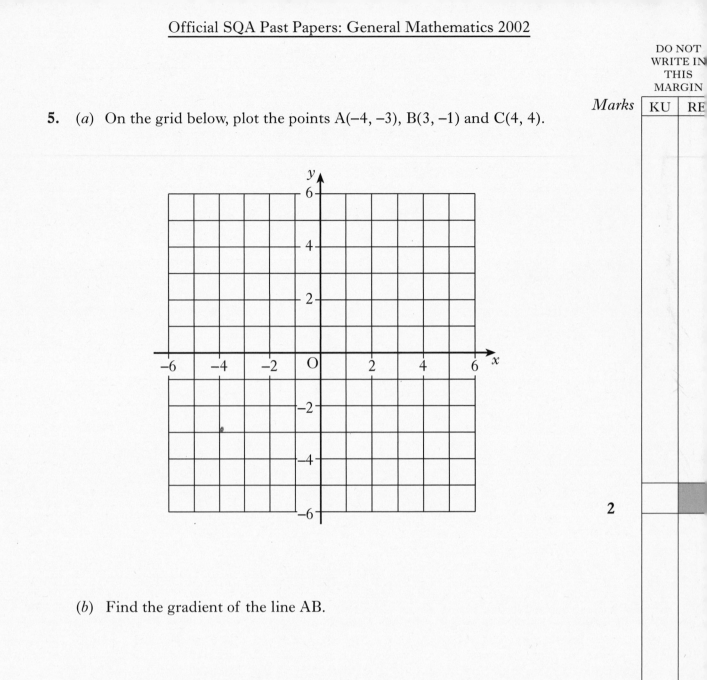

2

(*b*) Find the gradient of the line AB.

2

(*c*) Plot the fourth point D so that shape ABCD is a parallelogram.

Write down the coordinates of point D.

2

6. Starting with the smallest, write the following in order.

$$0{\cdot}404 \qquad \frac{1}{4} \qquad 41\% \qquad 0{\cdot}04 \qquad \frac{4}{10}$$

2

7.

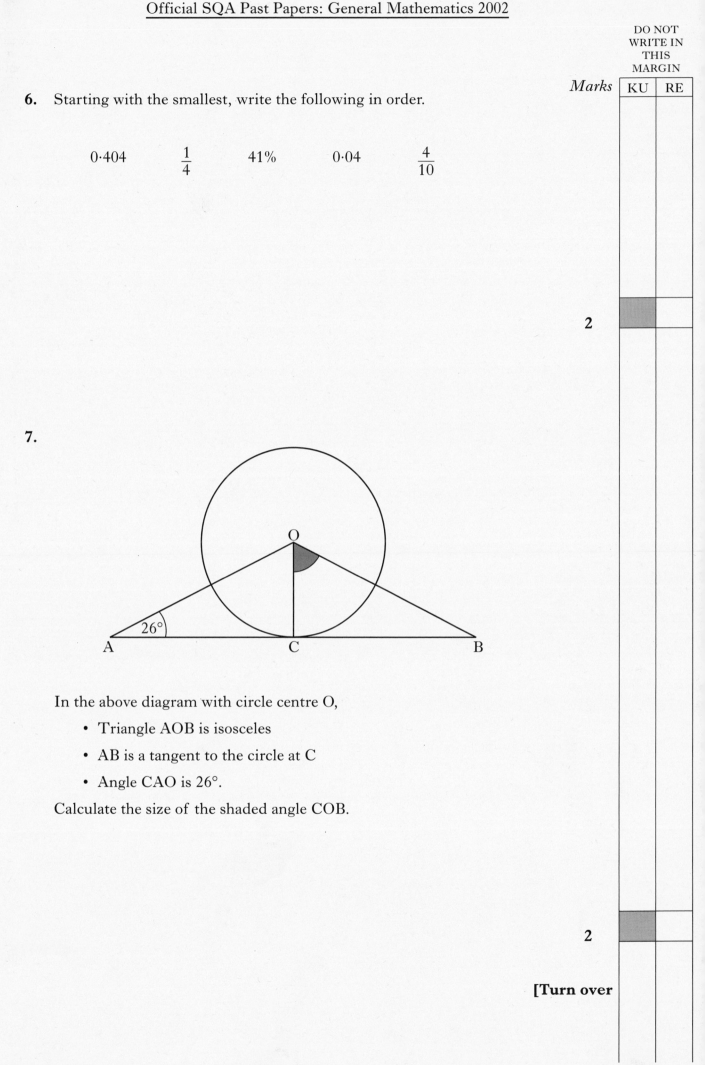

In the above diagram with circle centre O,

- Triangle AOB is isosceles
- AB is a tangent to the circle at C
- Angle CAO is 26°.

Calculate the size of the shaded angle COB.

2

[Turn over

8. The Science and Mathematics marks for 10 students are shown in the table below.

Student	A	B	C	D	E	F	G	H	I	J
Science mark	35	45	65	70	57	25	80	85	10	34
Mathematics mark	41	52	65	75	60	28	84	90	11	37

(*a*) Using these marks draw a Scattergraph.

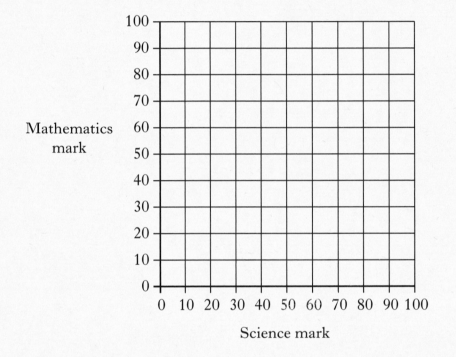

Mathematics mark

Science mark

3

(*b*) Draw a best-fitting line on the graph.

1

(*c*) A student whose Science mark is 50 was absent from the Mathematics exam.

Using the best-fitting line, estimate this student's Mathematics mark.

1

Marks | KU | RE

9. A gardener has been measuring the weekly growth rates of plants.

Two of the plants that have been measured are Plant A and Plant B.

One week Plant A is 29 cm high and Plant B is 46 cm high.

The next week Plant A is 57 cm high and Plant B is 71 cm high.

Which plant has grown more in the week and by how much?

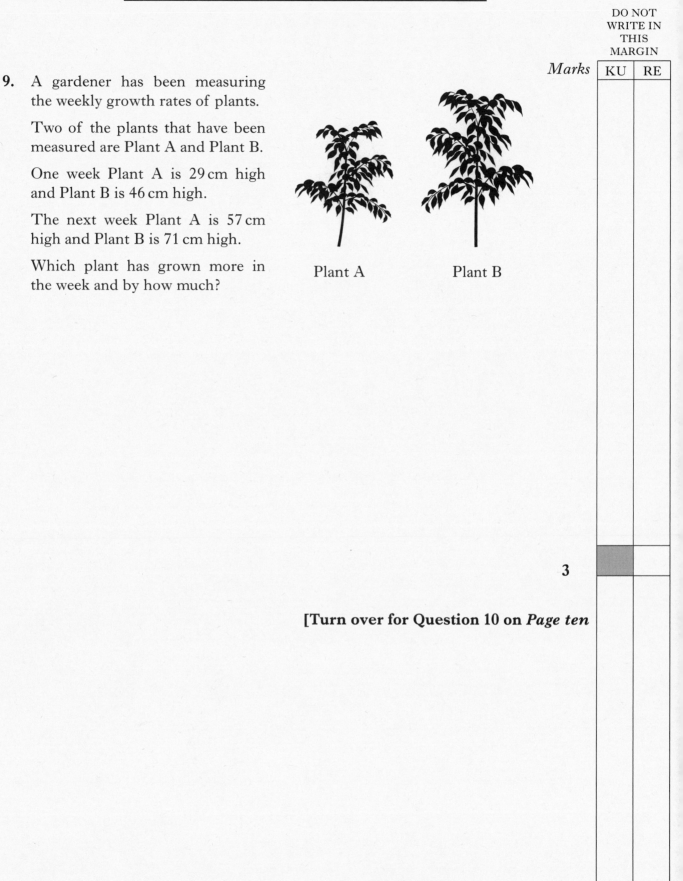

Plant A   Plant B

3

[Turn over for Question 10 on *Page ten*

10.

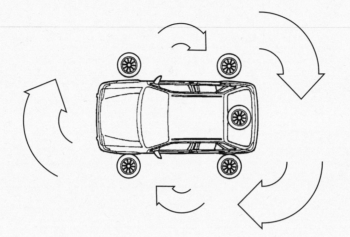

A car has five tyres, one on each of the four road wheels and one on the spare wheel.

Mr Anderson switched his wheels regularly so that all five tyres were used equally.

Last year he travelled 20 000 miles.

How many miles did each tyre do on the road?

3

[END OF QUESTION PAPER]

FOR OFFICIAL USE

G

KU   RE

Total marks

2500/404

NATIONAL
QUALIFICATIONS
2002

THURSDAY, 9 MAY
11.35 AM – 12.30 PM

MATHEMATICS
STANDARD GRADE
General Level
Paper 2

Fill in these boxes and read what is printed below.

Full name of centre

Town

Forename(s)

Surname

Date of birth
Day Month Year   Scottish candidate number   Number of seat

1 **You may use a calculator.**

2 Answer as many questions as you can.

3 Write your working and answers in the spaces provided. Additional space is provided at the end of this question-answer book for use if required. If you use this space, write clearly the number of the question involved.

4 Full credit will be given only where the solution contains appropriate working.

5 Before leaving the examination room you must give this book to the invigilator. If you do not you may lose all the marks for this paper.

SCOTTISH
QUALIFICATIONS
AUTHORITY

FORMULAE LIST

Circumference of a circle: $C = \pi d$

Area of a circle: $A = \pi r^2$

Curved surface area of a cylinder: $A = 2\pi rh$

Volume of a cylinder: $V = \pi r^2 h$

Volume of a triangular prism: $V = Ah$

Theorem of Pythagoras:

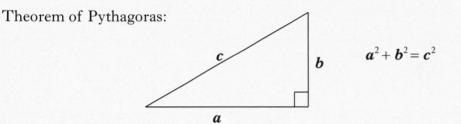

$$a^2 + b^2 = c^2$$

Trigonometric ratios
in a right angled
triangle:

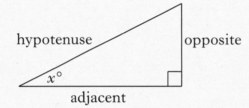
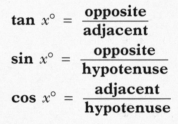

$$\tan x^\circ = \frac{\text{opposite}}{\text{adjacent}}$$

$$\sin x^\circ = \frac{\text{opposite}}{\text{hypotenuse}}$$

$$\cos x^\circ = \frac{\text{adjacent}}{\text{hypotenuse}}$$

Gradient:

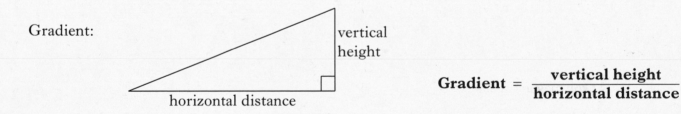

$$\text{Gradient} = \frac{\text{vertical height}}{\text{horizontal distance}}$$

DO NOT
WRITE IN
THIS
MARGIN

Marks KU | RE

1. John drives from Edinburgh to Inverness at an average speed of 76 kilometres per hour.

The journey takes him 3 hours 45 minutes.

How far is it from Edinburgh to Inverness?

2

[Turn over

Marks | KU | RF

2. Andrea sees this advertisement for a computer in CompCo.

CompCo
SPECIAL OFFER
£779 + VAT (17·5%)

OUR PROMISE
If you find the same computer at a cheaper price within 1 month, we will **refund double the difference**.

(a) Andrea buys the computer from CompCo.

VAT is 17·5%.

What is the total cost of the computer?

Round your answer to the nearest penny.

3

(b) One week later, Andrea sees the same computer in a different shop at £900 including VAT.

She remembers the promise in the CompCo advertisement and returns to the shop to claim a refund.

How much money should be refunded to her?

2

3. A column is in the shape of a cylinder.

It is 450 centimetres high and its diameter is 40 centimetres.

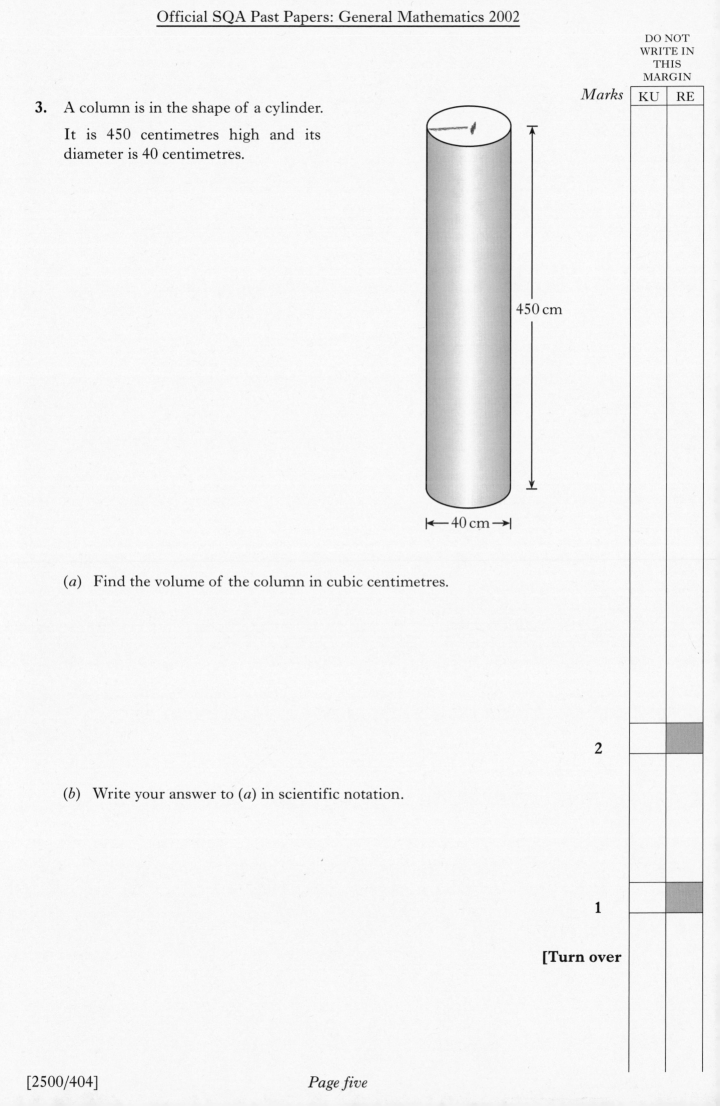

450 cm

|←—40 cm—→|

(a) Find the volume of the column in cubic centimetres.

2

(b) Write your answer to (a) in scientific notation.

1

[Turn over

DO NOT
WRITE I
THIS
MARGIN

Marks | KU | R

4. A metal fence for a garden is made by joining iron bars as shown below.

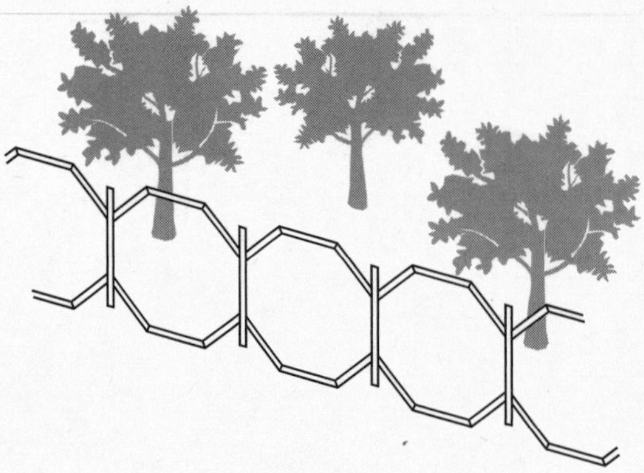

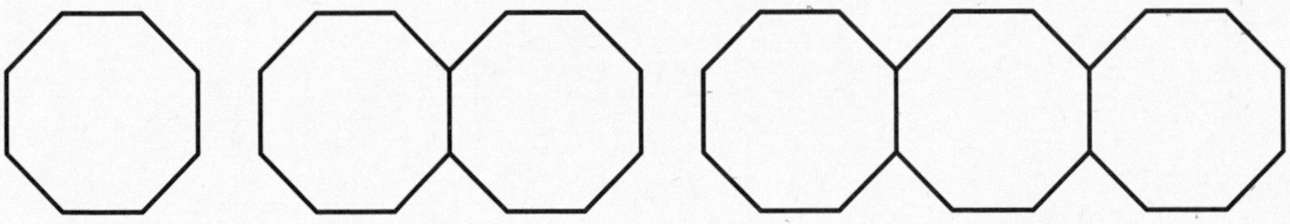

1 Section   2 Sections   3 Sections

(a) Complete this table.

Number of sections (s)	1	2	3	4		12
Number of iron bars (b)	8		22			

2

(b) Find a formula for calculating the number of iron bars (b), when you know the number of sections (s).

2

(c) A fence has been made by joining 176 iron bars.

How many sections are in this fence?

2

DO NOT WRITE IN THIS MARGIN

Marks	KU	RE

5. A sum of £1640 is invested in a bank.

The rate of interest is 4·5% per annum.

Calculate the simple interest gained in 9 months.

3

[Turn over

Marks | KU | R

6. PQRS is a rhombus.

Its diagonals PR and SQ are 20 centimetres and 12 centimetres long respectively.

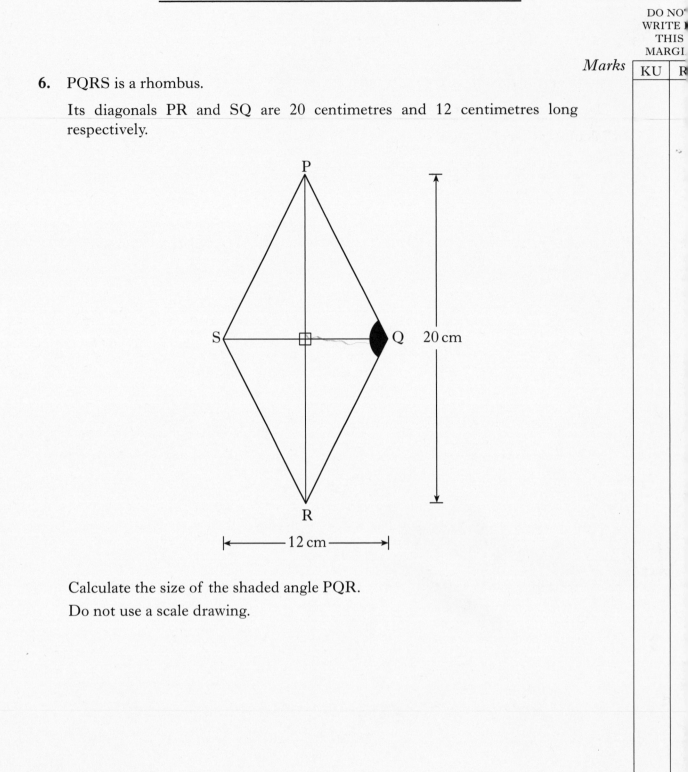

Calculate the size of the shaded angle PQR.

Do not use a scale drawing.

4

DO NOT WRITE IN THIS MARGIN

Marks | KU | RE

7. The diagram below shows the wall Jamie has tiled above the bath in his house.

He used rectangular tiles, some of which he halved.

The length of each tile is 30 centimetres.

The breadth of each tile is 20 centimetres.

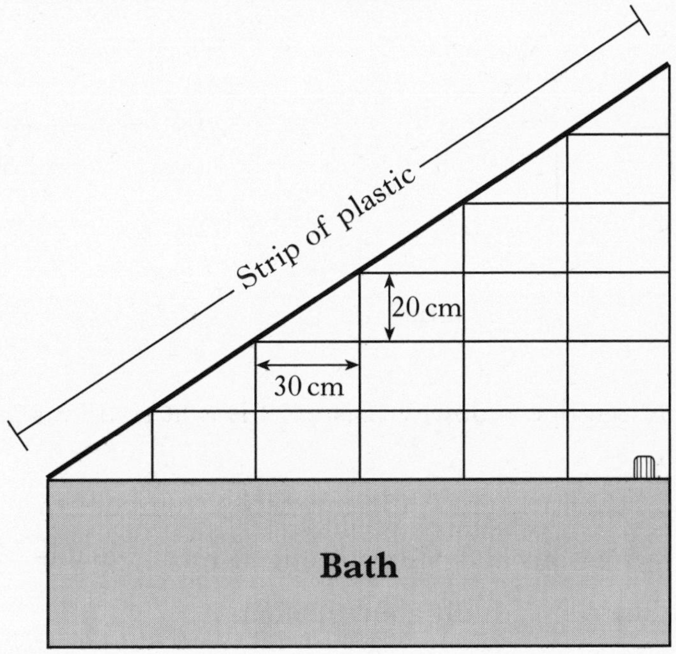

A strip of plastic is fitted along the top of the tiles.

Calculate the length of the strip of plastic.

4

[Turn over

DO NOT WRITE IN THIS MARGIN

Marks | KU | RE

8.

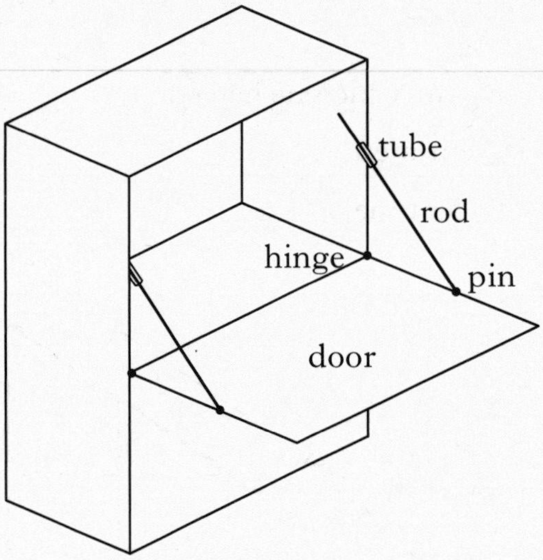

Diagram 1

A cabinet has a door that opens downwards until it is at right angles to the front of the cabinet.

A rod is pinned to the door at point P, 15 centimetres from the hinge, H.

The rod is 35 centimetres long and passes through a tube, at point T.

This tube is 20 centimetres vertically above the hinge.

8. (continued)

(*a*) Diagram 2 shows the positions of points P, T and H when the door is fully open.

Draw this diagram to a scale of 1:2.

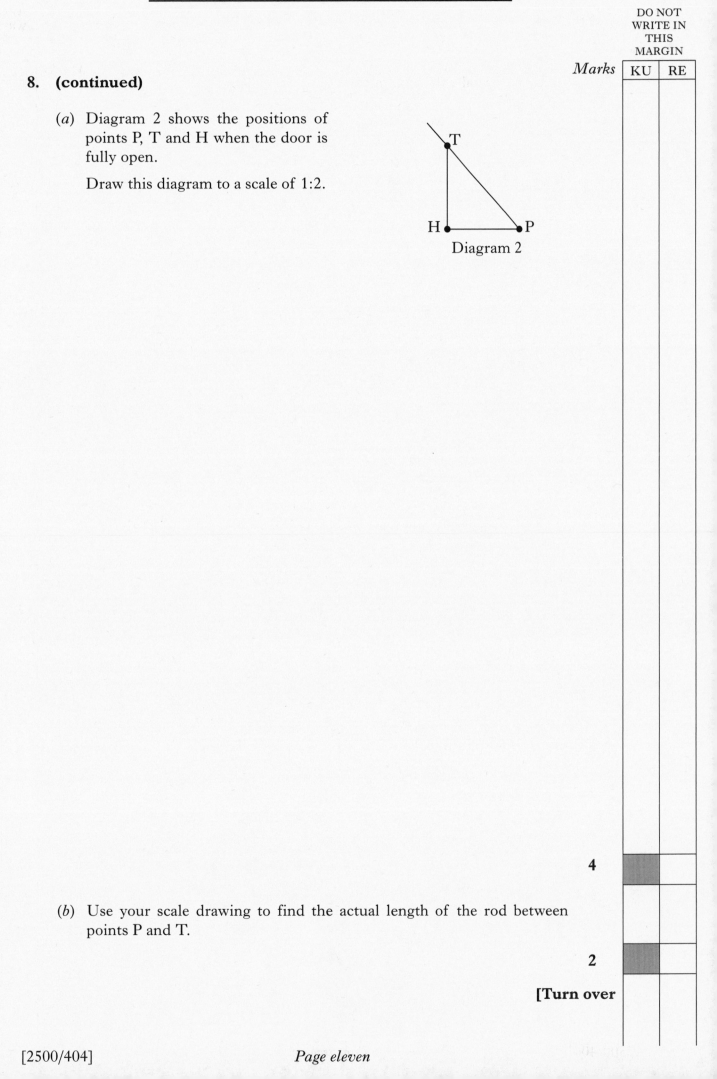

Diagram 2

4

(*b*) Use your scale drawing to find the actual length of the rod between points P and T.

2

[Turn over

Marks   KU

9. (*a*) Solve algebraically the equation

$$4(3x + 2) = 68.$$

3

(*b*) Factorise

$$10y + 15.$$

2

10. A joiner is making tables for a new coffee shop.

The shape of the top of a table is a semi-circle as shown below.

AB = 120 centimetres.

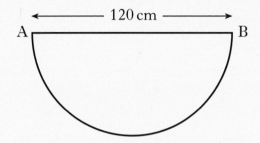

The top of the table is made of wood and a metal edge is to be fixed to its perimeter.

(*a*) Calculate the total length of the metal edge.

3

(*b*) The coffee shop needs 16 tables.

The joiner has 50 metres of the metal edge in the workshop.

Will this be enough for all sixteen tables?

Give a reason for your answer.

2

[Turn over

Marks

11. The Davidson family is planning to buy a new kitchen using hire purchase.

The cash price of the kitchen is £6300.

The hire purchase price is 22% more than the cash price.

The hire purchase agreement requires a deposit, which is 15% of the cash price, followed by 60 equal instalments.

Calculate the cost of each instalment.

4

[END OF QUESTION PAPER]

[BLANK PAGE]

FOR OFFICIAL USE

G

	KU	RE
Total marks		

2500/403

NATIONAL
QUALIFICATIONS
2003

THURSDAY, 8 MAY
10.40 AM – 11.15 AM

MATHEMATICS
STANDARD GRADE
General Level
Paper 1
Non-calculator

Fill in these boxes and read what is printed below.

Full name of centre

Town

Forename(s)

Surname

Date of birth
Day Month Year   Scottish candidate number   Number of seat

1 **You may not use a calculator.**

2 Answer as many questions as you can.

3 Write your working and answers in the spaces provided. Additional space is provided at the end of this question-answer book for use if required. If you use this space, write clearly the number of the question involved.

4 Full credit will be given only where the solution contains appropriate working.

5 Before leaving the examination room you must give this book to the invigilator. If you do not you may lose all the marks for this paper.

SCOTTISH
QUALIFICATIONS
AUTHORITY

©

FORMULAE LIST

Circumference of a circle: $C = \pi d$

Area of a circle: $A = \pi r^2$

Curved surface area of a cylinder: $A = 2\pi rh$

Volume of a cylinder: $V = \pi r^2 h$

Volume of a triangular prism: $V = Ah$

Theorem of Pythagoras:

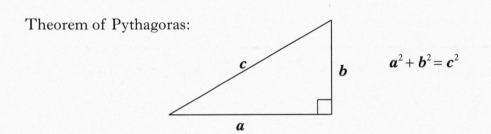

$$a^2 + b^2 = c^2$$

Trigonometric ratios
in a right angled
triangle:

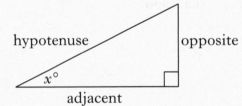

$$\tan x° = \frac{\text{opposite}}{\text{adjacent}}$$

$$\sin x° = \frac{\text{opposite}}{\text{hypotenuse}}$$

$$\cos x° = \frac{\text{adjacent}}{\text{hypotenuse}}$$

Gradient:

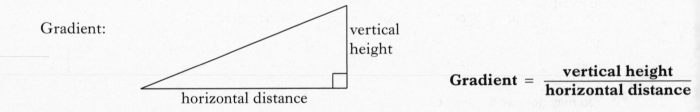

$$\text{Gradient} = \frac{\text{vertical height}}{\text{horizontal distance}}$$

DO NOT
WRITE IN
THIS
MARGIN

Marks | KU | RE

1. Carry out the following calculations.

(*a*) $3 \cdot 58 - 2 \cdot 734$

1

(*b*) $6 \cdot 37 \times 60$

1

(*c*) $13 \cdot 8 \div 4$

1

(*d*) $\frac{3}{4} + \frac{1}{16}$

2

[Turn over

2. Bruce sets out from base during an orienteering competition.

The arrow in the sketch below shows the direction in which he is travelling.

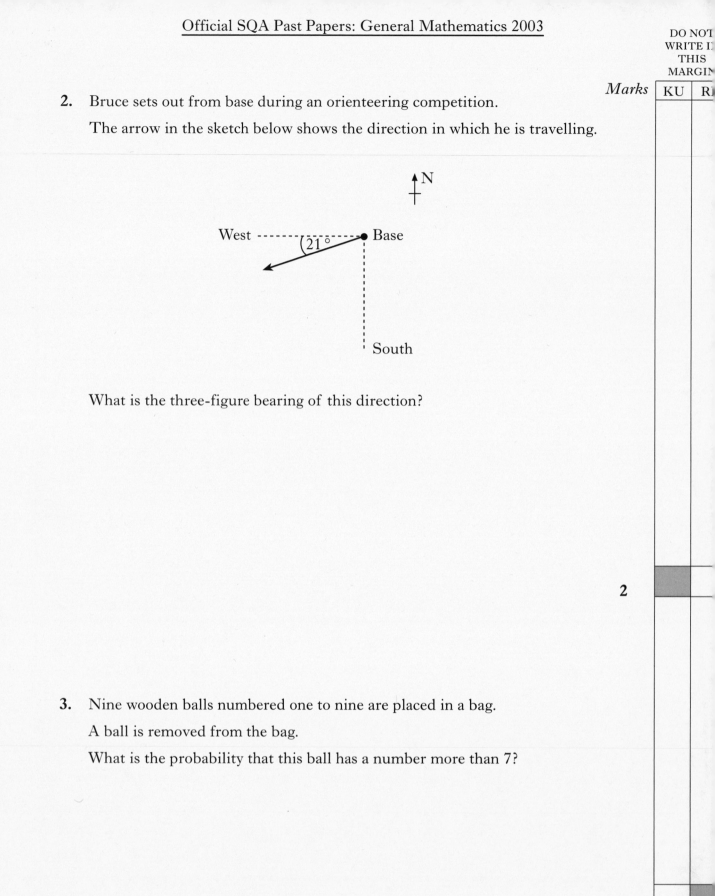

What is the three-figure bearing of this direction?

2

3. Nine wooden balls numbered one to nine are placed in a bag.

A ball is removed from the bag.

What is the probability that this ball has a number more than 7?

2

4. The letter A is shown in the diagram.

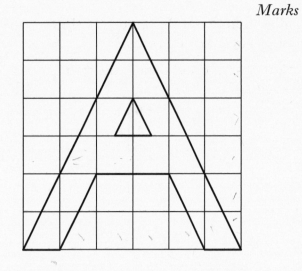

On the grid below, draw an enlargement of this letter A using a scale factor of 2.

3

[Turn over

DO NO'
WRITE
THIS
MARGI

Marks   KU   R

5. The number of hours of sunshine was recorded daily in a city during a three-week period in June.

The results are shown in the stem and leaf diagram below.

```
0 | 8
1 |
2 | 1  3
3 | 2  5  7
4 | 1  5  7  8
5 | 2  3  6
6 | 0  2  2
7 | 1  1  3  7  9
```

n = 21   3 | 2   represents 3·2 hours

Using the above diagram:

(a) calculate the range;

2

(b) find the median number of hours.

1

Marks | KU | RE

6. Four friends have dinner in a restaurant.

A service charge of 15% is added to their bill.

Their bill is shown below.

Armando's Pizza Restaurant

4 medium Pizzas   £ 28·00
4 large Colas   £ 5·00
     £ 33·00

Total including 15% service charge   £ 38·95

One of the friends thinks the service charge has been calculated wrongly.

Is the service charge correct?

Give a reason for your answer.

4

[Turn over

7. In a True or False game, players score +3 for a correct answer and −1 for a wrong answer.

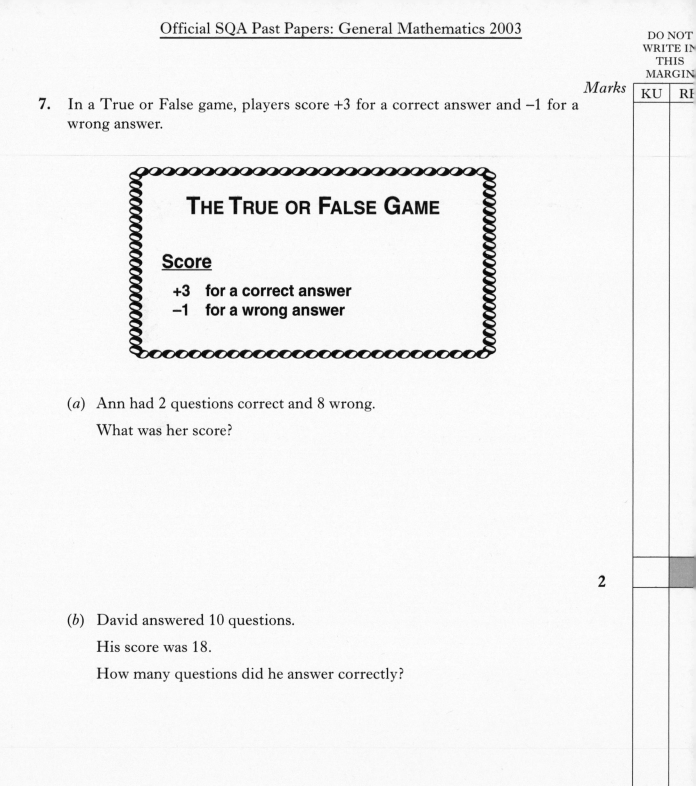

THE TRUE OR FALSE GAME

Score

+3   for a correct answer
−1   for a wrong answer

(a) Ann had 2 questions correct and 8 wrong.

What was her score?

2

(b) David answered 10 questions.

His score was 18.

How many questions did he answer correctly?

2

DO NOT WRITE IN THIS MARGIN

Marks   KU   RE

8. The international sizes for writing paper are shown in the list below.

All measurements are in millimetres.

A3	297	×	420
A4	210	×	297
A5	148	×	210
A6	105	×	148
A7	74	×	105
A8	52	×	74
A9	37	×	52
A10	26	×	37

By inspecting the list, write down the measurements for A10 writing paper.

2

9. The planet Pluto is approximately 7364 million kilometres from the Sun.

Write this number in scientific notation.

2

[Turn over for Question 10 on *Page ten*

10.

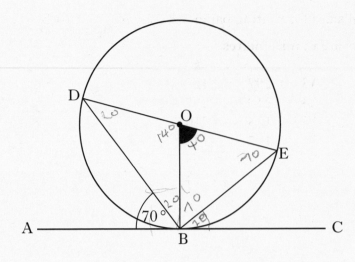

In the diagram above

- a circle, centre O, is drawn,
- the line AC is a tangent to the circle at B,
- Angle DBA = 70°.

Calculate the size of the shaded angle BOE.

3

[END OF QUESTION PAPER]

FOR OFFICIAL USE

G

Total marks

KU	RE

2500/404

NATIONAL QUALIFICATIONS 2003

THURSDAY, 8 MAY 11.35 AM – 12.30 PM

MATHEMATICS
STANDARD GRADE
General Level
Paper 2

Fill in these boxes and read what is printed below.

Full name of centre

Town

Forename(s)

Surname

Date of birth
Day Month Year

Scottish candidate number

Number of seat

1 **You may use a calculator.**

2 Answer as many questions as you can.

3 Write your working and answers in the spaces provided. Additional space is provided at the end of this question-answer book for use if required. If you use this space, write clearly the number of the question involved.

4 Full credit will be given only where the solution contains appropriate working.

5 Before leaving the examination room you must give this book to the invigilator. If you do not you may lose all the marks for this paper.

SCOTTISH
QUALIFICATIONS
AUTHORITY

©

FORMULAE LIST

Circumference of a circle: $C = \pi d$

Area of a circle: $A = \pi r^2$

Curved surface area of a cylinder: $A = 2\pi rh$

Volume of a cylinder: $V = \pi r^2 h$

Volume of a triangular prism: $V = Ah$

Theorem of Pythagoras:

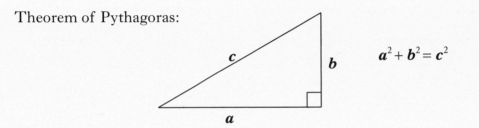

$$a^2 + b^2 = c^2$$

Trigonometric ratios
in a right angled
triangle:

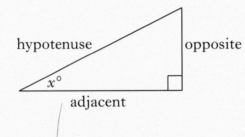

$$\tan x^\circ = \frac{\text{opposite}}{\text{adjacent}}$$

$$\sin x^\circ = \frac{\text{opposite}}{\text{hypotenuse}}$$

$$\cos x^\circ = \frac{\text{adjacent}}{\text{hypotenuse}}$$

Gradient:

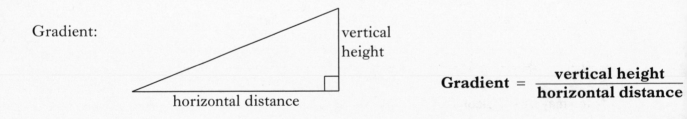

$$\text{Gradient} = \frac{\text{vertical height}}{\text{horizontal distance}}$$

1. The distance between Verona and Milan is 158 kilometres.

A train takes 1 hour 40 minutes to travel between these cities.

Find the average speed of the train.

2

2. Alice Anderson has a part-time job in a call centre.

Her basic rate of pay is £6·50 per hour.

At weekends she gets paid overtime at time and a half.

Last week she was paid £136·50, which included 4 hours overtime.

How many hours did she work at the basic rate?

4

[Turn over

Marks

3. The number of letters in each of the first one hundred words of a news story were counted.

The results are shown in the table below.

Number of letters	Frequency	Number of letters × frequency
1	5	
2	12	
3	18	
4	26	
5	18	
6	11	
7	7	
8	3	
Total =	Total =	

Find the mean number of letters per word.

Give your answer correct to one decimal place.

4

Official SQA Past Papers: General Mathematics 2003

DO NOT
WRITE IN
THIS
MARGIN

Marks | KU | RE

4.

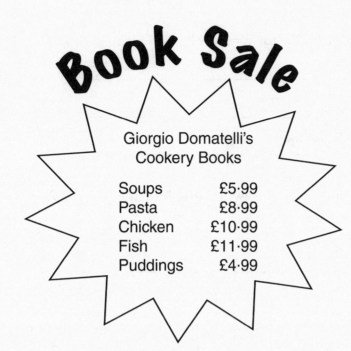

Giorgio Domatelli's
Cookery Books

Soups	£5·99
Pasta	£8·99
Chicken	£10·99
Fish	£11·99
Puddings	£4·99

Dayna wants to buy cookery books.

She chooses books from the cookery series shown above.

- She wants to spend between £15 and £20.
- She does not buy more than one copy of any book.

One way Dayna can choose her books is shown in the table below.

Complete the table to show all the different ways Dayna can choose her books.

BOOK TITLE	BOOK TITLE	BOOK TITLE	TOTAL COST (£)
Pasta	Chicken		19·98

3

[Turn over

DO NO
WRITE
THIS
MARGI

Marks | KU |

5. (*a*) Complete the table below for $y = 2x - 1$.

x	−4	0	4
y			

2

(*b*) Using the table in part (*a*), draw the graph of the line $y = 2x - 1$ on the grid below.

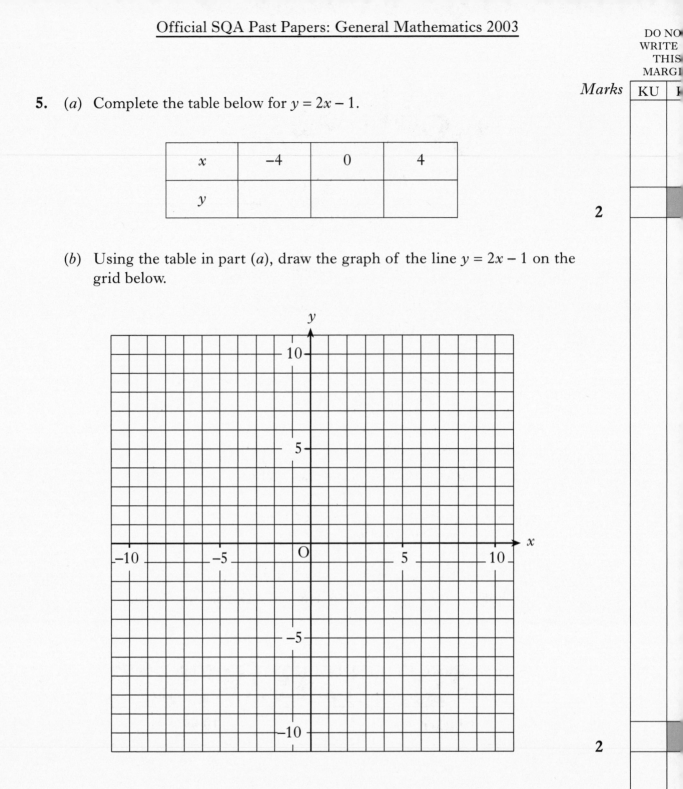

2

Marks | KU | RE

6.

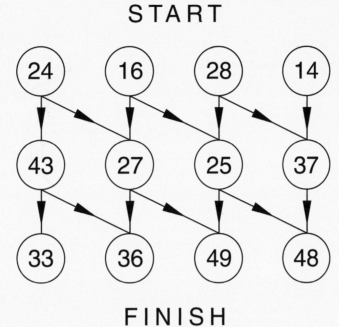

START

Following the arrows, use the instructions below.

Find the path which

- starts with a multiple of 4,
- moves to a prime number,
- finishes with a square number.

Write your numbers in the boxes below.

First number   Second number   Third number

3

[Turn over

7. The diagram shows the goal in American
Football and its shadow.

The post below the crossbar is 3 metres
high and casts a shadow 4 metres long.

The total length of the shadow is 9 metres.

Find the total goal height.

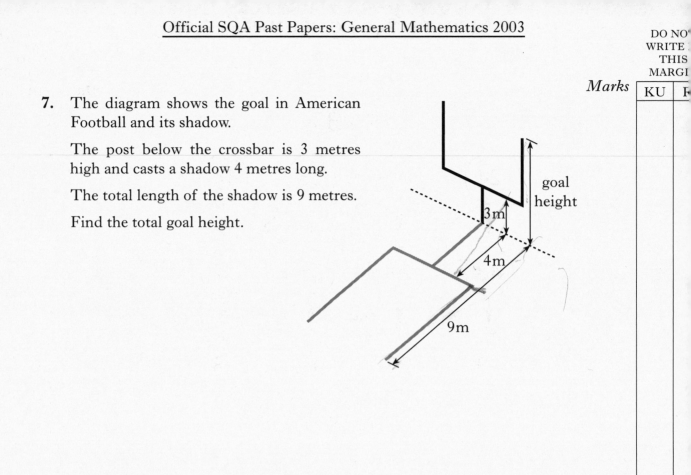

3

Marks | KU | RE

8. Alison has started a small business making wax candles.

She makes only one size of candle and it is in the shape of a cuboid.

The base of the candle is a square of side 6 centimetres.

The height of the candle is 15 centimetres.

Alison buys her wax in 10 litre tubs.

How many candles can she make from a tub of wax?

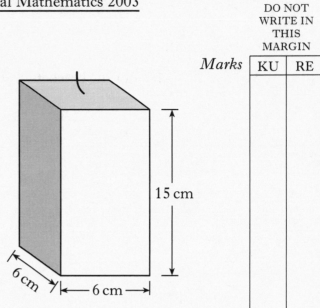

15 cm

6 cm

6 cm

4

[Turn over

Marks

9. (a) Multiply out the brackets and collect like terms

$$3(2w + 1) + 2(8 - w).$$

3

(b) Solve the inequality

$$3x - 4 < 11.$$

2

10. The cost, c pounds, of a carpet varies directly as its length, l metres. A carpet of length 5 metres costs £340.

(a) What will a carpet of length 8 metres cost?

3

(b) What length is a carpet which costs £238?

2

11. An adventure park is installing a climbing wall.

The wall is in the shape of a cylinder to which climbing pegs are attached.

The radius of the cylinder is 1·5 metres.

The cylinder has a curved surface area of 75·5 square metres.

What height will the cylinder be?

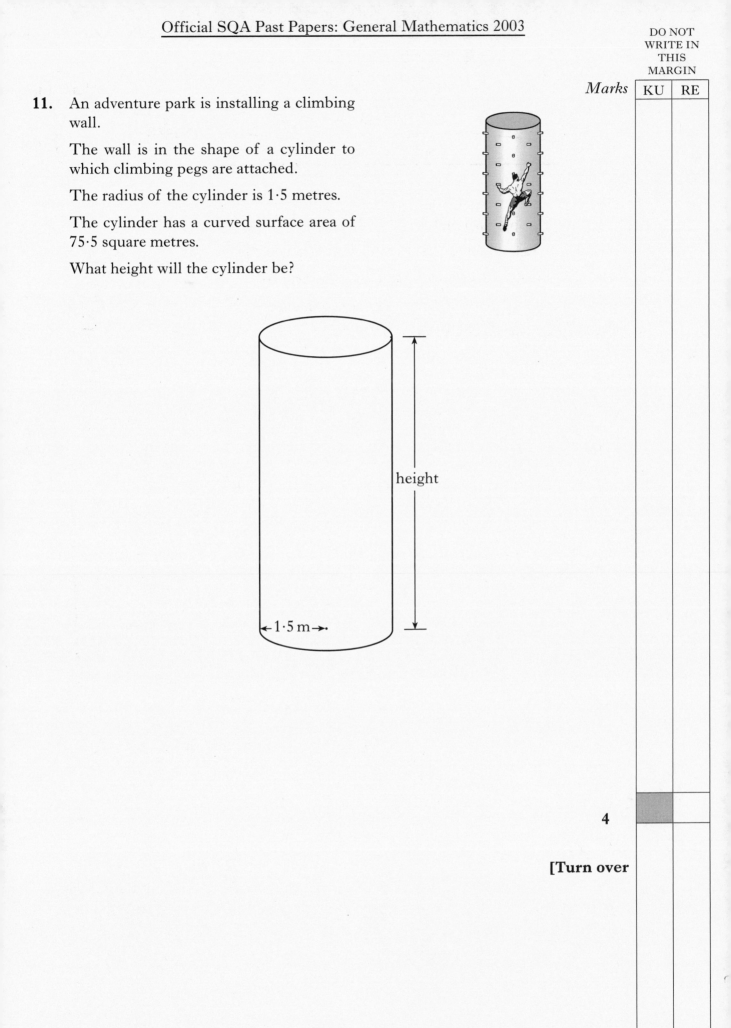

4

[Turn over

Page eleven

Marks | KU |

12.

5 km

Airport .... 7°)

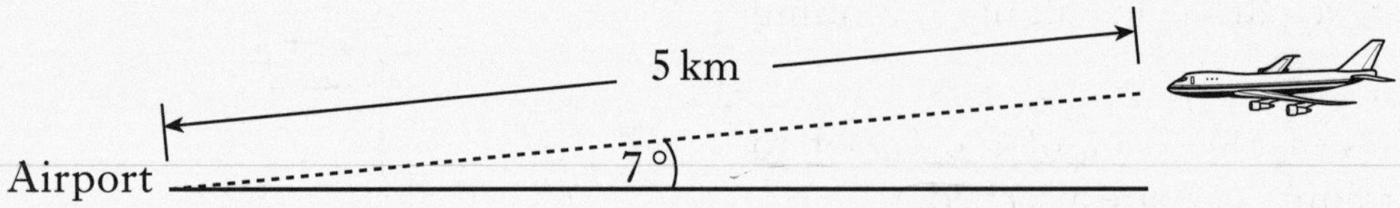

An aircraft is approaching Glasgow airport.

The angle of elevation of the aircraft from the airport is 7°.

The aircraft is at a distance of 5 km from the airport.

Find the height of the aircraft, to the nearest metre.

Do not use a scale drawing.

4

Page twelve

13. A large advertising banner is hanging from a building.

The banner is an isosceles triangle.

The top edge of the banner is 20 metres long and each of the other two sides is 26 metres long.

Find the area of the banner.

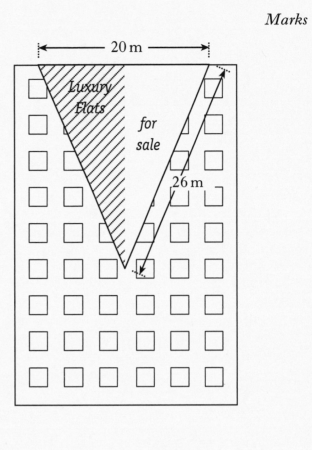

4

[END OF QUESTION PAPER]

[BLANK PAGE]

[BLANK PAGE]

FOR OFFICIAL USE

G

Total marks

KU	RE

2500/403

NATIONAL
QUALIFICATIONS
2004

FRIDAY, 7 MAY
10.40 AM – 11.15 AM

MATHEMATICS
STANDARD GRADE
General Level
Paper 1
Non-calculator

Fill in these boxes and read what is printed below.

Full name of centre

Town

Forename(s)

Surname

Date of birth

Day Month Year   Scottish candidate number   Number of seat

1 **You may not use a calculator.**

2 Answer as many questions as you can.

3 Write your working and answers in the spaces provided. Additional space is provided at the end of this question-answer book for use if required. If you use this space, write clearly the number of the question involved.

4 Full credit will be given only where the solution contains appropriate working.

5 Before leaving the examination room you must give this book to the invigilator. If you do not you may lose all the marks for this paper.

SCOTTISH
QUALIFICATIONS
AUTHORITY

©

FORMULAE LIST

Circumference of a circle: $C = \pi d$

Area of a circle: $A = \pi r^2$

Curved surface area of a cylinder: $A = 2\pi rh$

Volume of a cylinder: $V = \pi r^2 h$

Volume of a triangular prism: $V = Ah$

Theorem of Pythagoras:

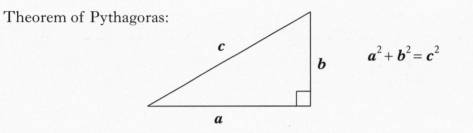

$$a^2 + b^2 = c^2$$

Trigonometric ratios
in a right angled
triangle:

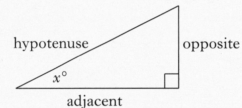 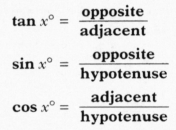

$$\tan x^\circ = \frac{\text{opposite}}{\text{adjacent}}$$

$$\sin x^\circ = \frac{\text{opposite}}{\text{hypotenuse}}$$

$$\cos x^\circ = \frac{\text{adjacent}}{\text{hypotenuse}}$$

Gradient:

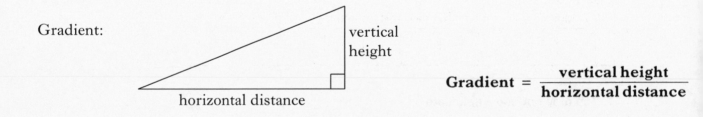

$$\textbf{Gradient} = \frac{\textbf{vertical height}}{\textbf{horizontal distance}}$$

DO NOT
WRITE IN
THIS
MARGIN

Marks KU RE

1. Carry out the following calculations.

(*a*) $14 \cdot 93 - 3 \cdot 7 + 2 \cdot 15$

1

(*b*) $42 \cdot 8 \times 7$

1

(*c*) $1710 \div 3000$

1

(*d*) 90% of £180

2

2. Express $\frac{3}{7}$ as a decimal.

Give your answer correct to two decimal places.

2

Marks | KU | R

3. Ann Fiona Johnstone has drawn a design which uses her initials.

She wants her finished design to be symmetrical.

Complete her design so that the dotted line is an axis of symmetry.

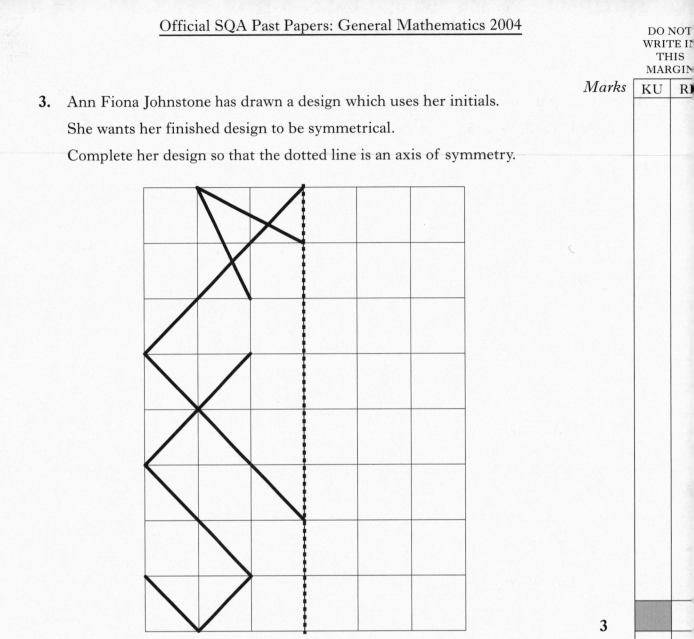

3

4. The largest ocean in the world is the Pacific Ocean.

Its area is approximately $1 \cdot 813 \times 10^8$ square kilometres.

Write this number in full.

1

Marks | KU | RE

5. A patient in hospital had his temperature checked every two hours.

The results are shown in the table below.

Time	12 noon	2 pm	4 pm	6 pm	8 pm	10 pm
Temperature (°C)	38·2	38·6	38·1	37·9	37·5	36·9

Illustrate this data on the grid below using a line graph.

Temperature Chart

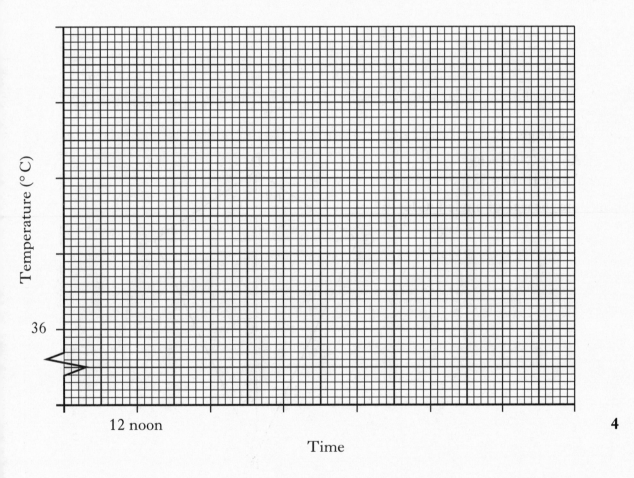

4

[Turn over

Page five

DO NOT
WRITE I?
THIS
MARGIN

Marks   KU   R?

6. Last month a garage sold 12 red cars, 9 silver cars and 15 black cars.

Joe bought one of these cars.

What is the probability that the car Joe bought was silver?

Give your answer as a fraction in its simplest form.

2

7. DEFG is a kite.

- Angle GDF = 69°
- Angle EFD = 33°

Calculate the size of angle DGF.

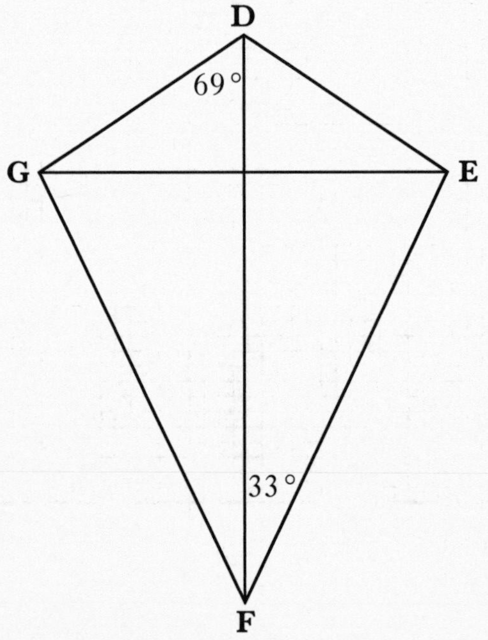

3

Marks KU RE

8. Christy needs a four-digit code to switch on her mobile phone.

She uses the digits from her birth date 4/3/89, but in a different order.

She knows that the last digit is 8.

One of the possible four-digit codes Christy could try is shown in the table below.

Complete the table to show all the possible four-digit codes.

4	3	9	8

3

9. A recipe for Shortbread uses the following ingredients.

  300 grams flour
  100 grams sugar
  200 grams butter

Alana has only 240 grams of flour.

To make Shortbread using all of the 240 grams of flour she will have to adjust the quantities of sugar and butter.

How many grams of sugar and how many grams of butter should she use?

4

Marks

10. The heating in Bruce's house switches on automatically when the outside temperature drops to –5 °C.

One day last winter the outside temperature was 3 °C.

Calculate the drop in temperature when the heating switched on automatically.

2

11. Andrew and his brother are flying to America on holiday.

Their flight times are shown below.

Depart Glasgow	30/6/04	2120
Arrive Reykjavik, Iceland	30/6/04	2235
Depart Reykjavik, Iceland	1/7/04	0105
Arrive New York, USA	1/7/04	0455

How long will the brothers have to wait at Reykjavik in Iceland before their flight to New York?

2

[END OF QUESTION PAPER]

FOR OFFICIAL USE

G

Total marks

KU	RE

2500/404

NATIONAL
QUALIFICATIONS
2004

FRIDAY, 7 MAY
11.35 AM – 12.30 PM

MATHEMATICS
STANDARD GRADE
General Level
Paper 2

Fill in these boxes and read what is printed below.

Full name of centre

Town

Forename(s)

Surname

Date of birth
Day Month Year   Scottish candidate number   Number of seat

1 **You may use a calculator.**

2 Answer as many questions as you can.

3 Write your working and answers in the spaces provided. Additional space is provided at the end of this question-answer book for use if required. If you use this space, write clearly the number of the question involved.

4 Full credit will be given only where the solution contains appropriate working.

5 Before leaving the examination room you must give this book to the invigilator. If you do not you may lose all the marks for this paper.

SCOTTISH
QUALIFICATIONS
AUTHORITY

FORMULAE LIST

Circumference of a circle: $C = \pi d$

Area of a circle: $A = \pi r^2$

Curved surface area of a cylinder: $A = 2\pi rh$

Volume of a cylinder: $V = \pi r^2 h$

Volume of a triangular prism: $V = Ah$

Theorem of Pythagoras:

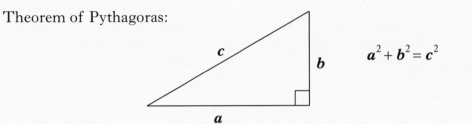

$$a^2 + b^2 = c^2$$

Trigonometric ratios
in a right angled
triangle:

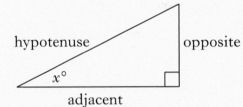

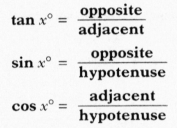

$$\tan x^\circ = \frac{\text{opposite}}{\text{adjacent}}$$

$$\sin x^\circ = \frac{\text{opposite}}{\text{hypotenuse}}$$

$$\cos x^\circ = \frac{\text{adjacent}}{\text{hypotenuse}}$$

Gradient:

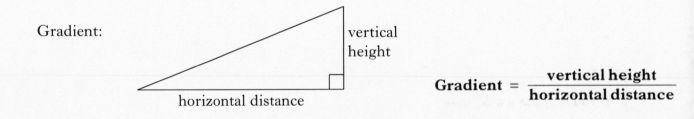

$$\text{Gradient} = \frac{\text{vertical height}}{\text{horizontal distance}}$$

DO NOT WRITE IN THIS MARGIN

Marks KU RE

1. 100 grams of wholemeal bread contain the following:

Protein	10 grams
Carbohydrates	55 grams
Fibre	9 grams
Fat	3 grams
Other	23 grams

A pie chart is to be drawn to show this information.

What size of angle should be used for the carbohydrates?

DO NOT DRAW A PIE CHART. 2

[Turn over

2. A company manufactures boxes of tacks and claims that there are "on average" 60 tacks per box.

This claim is tested by counting the number of tacks in a sample of 100 boxes.

The results are shown below.

Number of tacks	Frequency	Number of tacks × Frequency
57	7	
58	13	
59	21	
60	24	
61	19	
62	12	
63	4	
Totals	100	

(a) Find the mean number of tacks per box.

3

(b) Is the company's claim reasonable?

You must give a reason for your answer.

1

Marks

KU	RE

3. The sketch below shows the net of a three-dimensional shape.

The net consists of a rectangle and two equal circles of radius 3 centimetres.

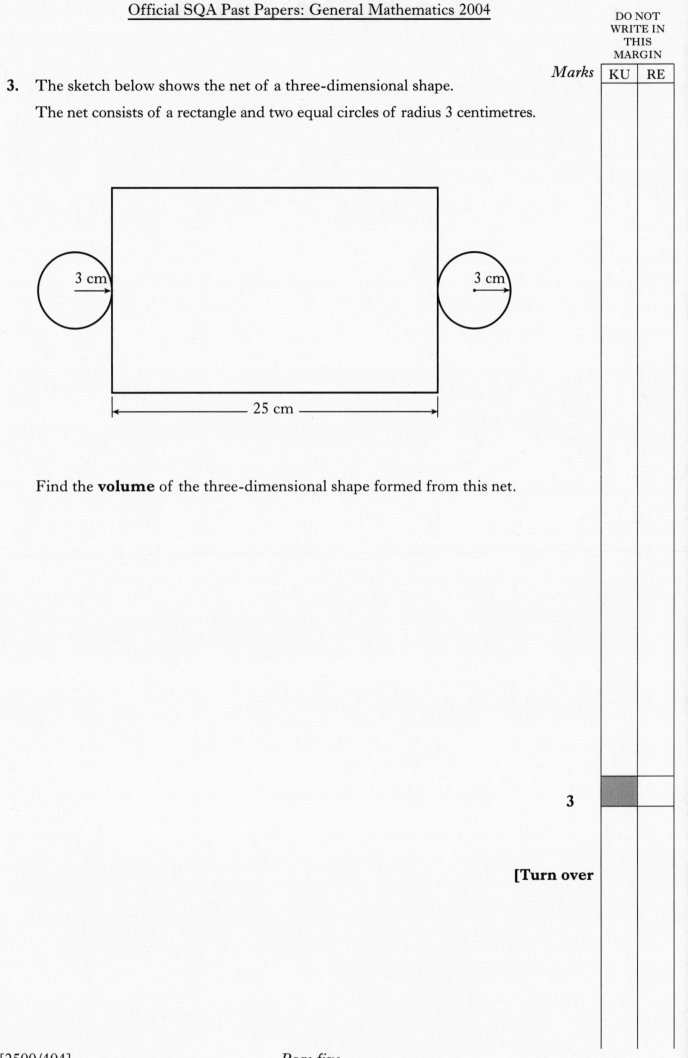

Find the **volume** of the three-dimensional shape formed from this net.

3

[Turn over

Page five

DO NOT WRITE IN THIS MARGIN

Marks | KU | RE

4. (*a*) Solve algebraically

$$5x - 2 = 2x + 19.$$

3

(*b*) Factorise fully

$$12 + 8p.$$

2

5.

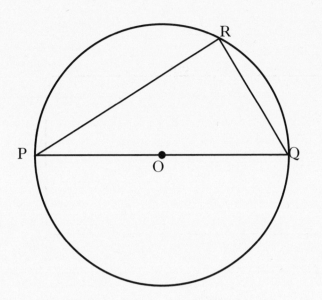

PQ is a diameter of the circle with centre O.

R is a point on the circumference of the circle.

PR is 12 centimetres.

RQ is 5·5 centimetres.

Calculate the length of the radius of the circle.

4

[Turn over

Marks

DO NOT
WRITE
THIS
MARGIN

KU

6.

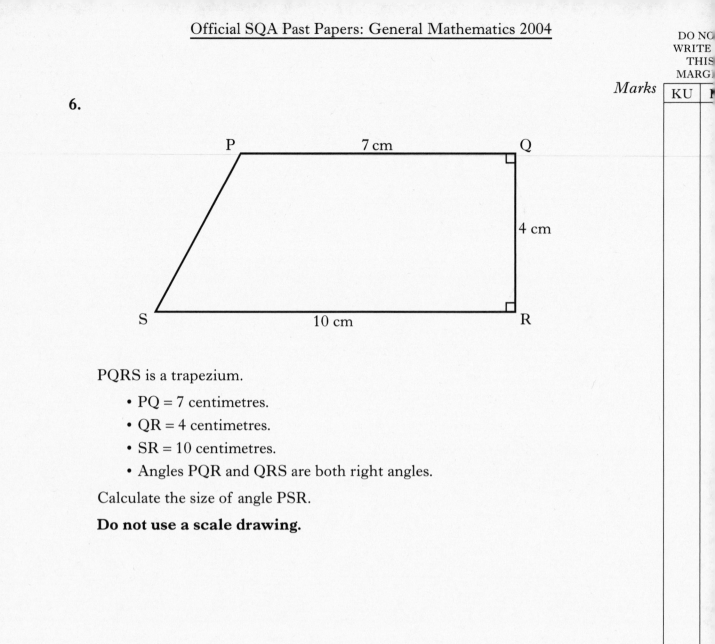

PQRS is a trapezium.

- PQ = 7 centimetres.
- QR = 4 centimetres.
- SR = 10 centimetres.
- Angles PQR and QRS are both right angles.

Calculate the size of angle PSR.

Do not use a scale drawing.

4

7. (*a*) John is going to Italy on holiday.

He changes £500 to Euros.

The exchange rate is £1 = 1·51 Euros.

How many Euros will he get?

2

(*b*) While in Italy he decides to visit Switzerland for a day.

He wants to change 100 Euros to Swiss Francs.

John knows the exchange rate is £1 = 2·33 Swiss Francs.

How many Swiss Francs should he get for 100 Euros?

3

KU | RE

Marks | KU | I

8. The floor of a conservatory consists of a rectangle and a semicircle.

The floor has the shape shown below.

The measurements are in metres.

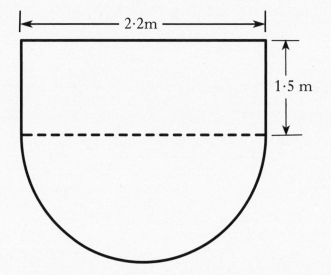

Find the total area of the floor.

4

Marks | KU | RE

9. A basic cable television package, which includes 30 channels, costs £8·75 per month.

The cost of installation is £75 which will be included in the first month's bill.

Additional channels can be added to the basic service.

- The movie channels package costs £12·50 per month.
- The music channels package costs £7·50 per month.
- The sports channels package costs £14·50 per month.

The Mackie family's first bill after having cable television installed was £98·25.

They chose the basic cable television package plus one additional channels package.

Which additional package did they choose?

Give a reason for your answer.

3

[Turn over

DO NO
WRITE
THIS
MARGI

Marks KU

10. Janice is planning to go on holiday to Florida.

   She wants to book with FloridaSun Holidays and stay at the Parkway Hotel.

   The table below shows the pricing information for the Parkway Hotel.

FloridaSun Holidays PARKWAY HOTEL	Prices are per person in £s	
No. of Nights	**7**	**14**
May 17	445	725
24	459	735
31	465	749
June 7	479	765
14	499	779
21	509	789
28	519	799
July 5	525	805
12	535	825
19	545	839
26	609	855
Aug 2	615	869
9	625	895
16	639	875
23	539	845
30	519	805

(a) Janice wants to stay in the Parkway Hotel for 14 nights.

   What will be the price if her holiday starts on 5th July?

1

Marks

10. **(continued)**

(b) The Parkway Hotel charges an extra £4·95 per person per night for a single room.

How much extra will Janice pay for her 14 night holiday if she wants a single room?

1

(c) If Janice books today she will get a 20% discount on her **total cost**.

Find the discounted price of her 14 night holiday in a single room from 5th July.

2

Marks KU

11. Mara travels 1850 miles every month.

Currently her car runs on unleaded petrol, which costs 76·9p per litre and her car travels 8·5 miles per litre.

(*a*) What is her monthly petrol bill?

2

Mara is thinking of having her car converted to run on Liquid Petroleum Gas (LPG).

LPG costs 38·9p per litre and using this fuel her car will travel 7·8 miles per litre.

(*b*) What will be her monthly saving if she converts her car to run on LPG?

2

11. (continued)

(c) The cost of converting Mara's car to run on LPG is £800.

How many months of savings will it take to recover the cost of the conversion?

2

[Turn over

DO NOT
WRITE
THIS
MARGI

Marks | KU | R

12. The current, C amps, of an electrical appliance is calculated using the formula

$$C = \frac{P}{240} \text{ , where } P \text{ watts is the power rating.}$$

• A hairdryer has a power rating of 850 watts.

• The fuse used should be the one just bigger than the calculated current.

• The choice of fuses is 3 amp, 5 amp and 13 amp.

Which fuse should be used?

3

13. The diagram below shows the position of two buoys.

Sofie has to sail her yacht between the two buoys so that it is always the same distance from each buoy.

Show the yacht's **course** on the diagram.

Buoy ⊕

⊕ **Buoy**

2

[END OF QUESTION PAPER]

ADDITIONAL SPACE FOR ANSWERS

[BLANK PAGE]

FOR OFFICIAL USE

G

Total marks

KU	RE

2500/403

NATIONAL
QUALIFICATIONS
2005

FRIDAY, 6 MAY
10.40 AM – 11.15 AM

MATHEMATICS
STANDARD GRADE
General Level
Paper 1
Non-calculator

Fill in these boxes and read what is printed below.

Full name of centre

Town

Forename(s)

Surname

Date of birth

Day Month Year   Scottish candidate number   Number of seat

1 You may **not** use a calculator.

2 Answer as many questions as you can.

3 Write your working and answers in the spaces provided. Additional space is provided at the end of this question-answer book for use if required. If you use this space, write clearly the number of the question involved.

4 Full credit will be given only where the solution contains appropriate working.

5 Before leaving the examination room you must give this book to the invigilator. If you do not you may lose all the marks for this paper.

SCOTTISH
QUALIFICATIONS
AUTHORITY

©

FORMULAE LIST

Circumference of a circle: $C = \pi d$

Area of a circle: $A = \pi r^2$

Curved surface area of a cylinder: $A = 2\pi rh$

Volume of a cylinder: $V = \pi r^2 h$

Volume of a triangular prism: $V = Ah$

Theorem of Pythagoras:

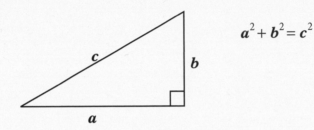

$$a^2 + b^2 = c^2$$

Trigonometric ratios
in a right angled
triangle:

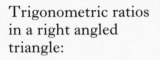

$$\tan x° = \frac{\text{opposite}}{\text{adjacent}}$$

$$\sin x° = \frac{\text{opposite}}{\text{hypotenuse}}$$

$$\cos x° = \frac{\text{adjacent}}{\text{hypotenuse}}$$

Gradient:

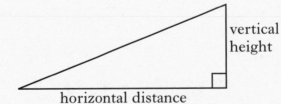

$$\text{Gradient} = \frac{\text{vertical height}}{\text{horizontal distance}}$$

Marks | KU | RE

1. Carry out the following calculations.

(a) $209 \cdot 3 - 175 \cdot 48$

1

(b) $56 \cdot 7 \times 90$

1

(c) $324 \cdot 1 \div 7$

1

(d) $\frac{3}{4}$ of 56 cm

2

2. When an aircraft leaves Prestwick airport the outside temperature is 12° Celsius.

The aircraft climbs to 10 000 metres and the outside temperature is −50° Celsius.

Find the difference between these temperatures.

KU | RE

2

3. Sandra is working on the design for a bracelet.

She is using matches to make each shape.

Shape 1 **Shape 2** **Shape 3** **Shape 4**

(a) Draw shape 4.

1

(b) Complete the following table.

Shape number (s)	1	2	3	4	5	6		13
Number of matches (m)	5	9			21			

2

(c) Find a formula for calculating the number of matches, (m), when you know the shape number, (s).

2

(d) Which shape number uses 61 matches?

You must show your working.

2

4. A ship is transporting 2800 cars.

Each car is worth £20 000.

(a) What is the total value of all the cars?

1

(b) Write the total value in scientific notation.

1

[Turn over

DO NOT
WRITE I
THIS
MARGIN

Marks KU R

5. (*a*) On the grid below, plot the points A(7, 5), B(5, –1) and C(–1, –3).

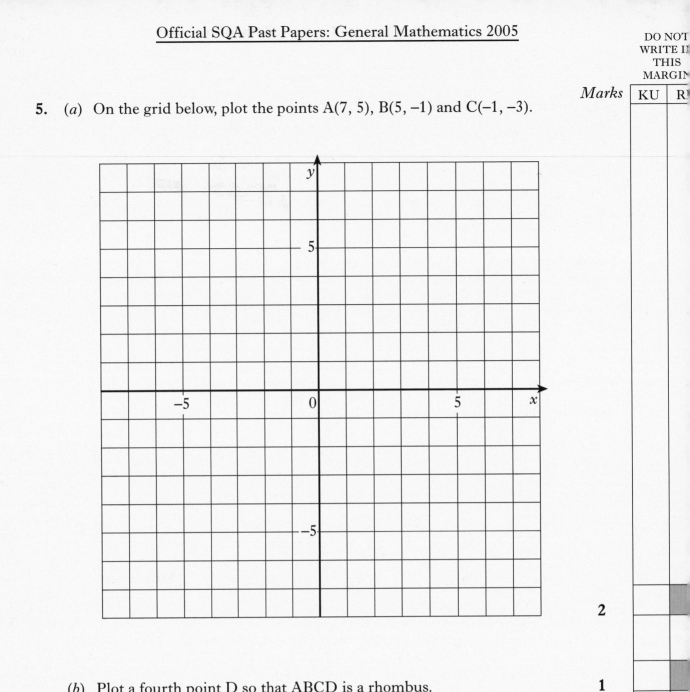

2

(*b*) Plot a fourth point D so that ABCD is a rhombus.

1

(*c*) Reflect rhombus ABCD in the **y-axis**.

2

6. The table below can be used to convert tyre pressures from pounds per square inch (lb/sq in) to kilograms per square centimetre (kg/sq cm).

lb/sq in	20	22	24	26	28	30	32	34
kg/sq cm	1·41	1·55	1·69	1·83	1·97	2·11	2·25	2·39

Convert **29 lb/sq in** to **kg/sq cm**.

2

7. (a) Graham goes into a shop and buys a bottle of water and a cheese roll for £1·38.

In the same shop, Alan pays £1·77 for 2 bottles of water and a cheese roll.

How much does a bottle of water cost?

1

(b) Craig goes into the shop and buys 4 bottles of water and 3 cheese rolls.

How much will this cost?

3

[Turn over

DO NOT WRITE IN THIS MARGIN

Marks | KU | RE

8. John buys a football programme for £1·60 and sells it for £2·00.

Calculate his percentage profit. **3**

9.

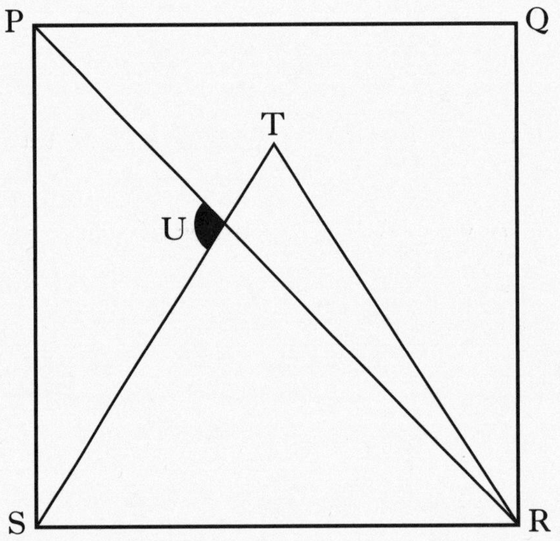

In the diagram above

- PQRS is a square
- PR is a diagonal of the square
- Triangle RST is equilateral.

Calculate the size of the shaded angle SUP. **3**

[*END OF QUESTION PAPER*]

ADDITIONAL SPACE FOR ANSWERS

ADDITIONAL SPACE FOR ANSWERS

ADDITIONAL SPACE FOR ANSWERS

[BLANK PAGE]

FOR OFFICIAL USE

G

KU   RE

Total marks

2500/404

NATIONAL
QUALIFICATIONS
2005

FRIDAY, 6 MAY
11.35 AM – 12.30 PM

MATHEMATICS
STANDARD GRADE
General Level
Paper 2

Fill in these boxes and read what is printed below.

Full name of centre

Town

Forename(s)

Surname

Date of birth

  Day Month Year   Scottish candidate number   Number of seat

1 **You may use a calculator.**

2 Answer as many questions as you can.

3 Write your working and answers in the spaces provided. Additional space is provided at the end of this question-answer book for use if required. If you use this space, write clearly the number of the question involved.

4 Full credit will be given only where the solution contains appropriate working.

5 Before leaving the examination room you must give this book to the invigilator. If you do not you may lose all the marks for this paper.

SCOTTISH
QUALIFICATIONS
AUTHORITY

©

FORMULAE LIST

Circumference of a circle:	$C = \pi d$
Area of a circle:	$A = \pi r^2$
Curved surface area of a cylinder:	$A = 2\pi rh$
Volume of a cylinder:	$V = \pi r^2 h$
Volume of a triangular prism:	$V = Ah$

Theorem of Pythagoras:

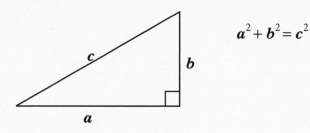

$$a^2 + b^2 = c^2$$

Trigonometric ratios
in a right angled
triangle:

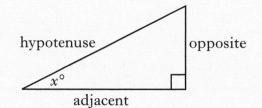

$$\tan x^\circ = \frac{\text{opposite}}{\text{adjacent}}$$

$$\sin x^\circ = \frac{\text{opposite}}{\text{hypotenuse}}$$

$$\cos x^\circ = \frac{\text{adjacent}}{\text{hypotenuse}}$$

Gradient:

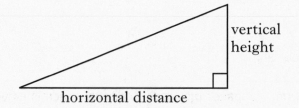

$$\text{Gradient} = \frac{\text{vertical height}}{\text{horizontal distance}}$$

DO NOT WRITE IN THIS MARGIN

Marks | KU | RE

1. A night train from London to Edinburgh leaves at 2321 and arrives at 0651.

(*a*) How long does the train journey take?

2

(*b*) The distance from London to Edinburgh is 644 kilometres.

Find the average speed of the train in kilometres per hour.

Give your answer correct to one decimal place.

3

[Turn over

DO NOT WRITE IN THIS MARGIN

Marks | KU | RE

2. The marks of a group of pupils in a maths test are shown below.

43	17	25	25	29	31	32	11	26	20
25	42	32	33	25	28	41	35	32	26

(*a*) Illustrate this data in an ordered stem and leaf diagram. **3**

(*b*) What is the mode for the above data? **1**

Marks | KU | RE

3. Scott sees the following notice in the window of the Big Computer Shop.

The Big Computer Shop

Massive Sale

$33\frac{1}{3}$% discount

on all purchases

(a) A computer was £834.

How much would Scott pay for it in the sale?

2

The same computer can be bought in Pete's PC Shop on hire purchase.

PETE'S PC SHOP

£55 deposit
and
£23·33 per month for 2 years

(b) Which shop sells the computer cheaper?

Show your working.

3

Marks

KU   R

4. The diagram below shows the shape of Sangita's garden.

Sangita plants a hedge along side AB.

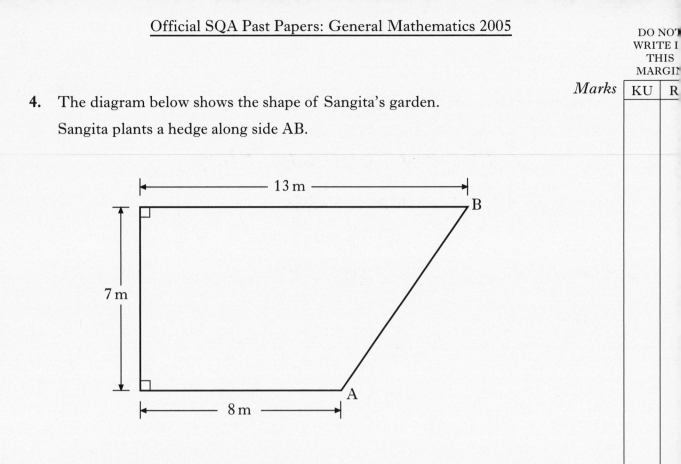

Calculate the length of the hedge.

4

Marks   KU   RE

5. (*a*) Remove the brackets and simplify

$$5 + 3(2x - 5).$$

2

(*b*) Solve the inequality

$$3x - 5 \geq 13.$$

2

[Turn over

DO NOT WRITE IN THIS MARGIN

Marks | KU | RE

6. The sponsors of the Champions league have given £900 000 to be shared among the four competing teams.

The league table is shown below.

The teams share the money in the ratio of the **points** they gain.

How much is **United's** share of the money?

	Played	Won	Lost	Drawn	Points
Inter	3	3	0	0	**9**
Athletic	3	2	1	0	**6**
United	3	1	2	0	**3**
Red Star	3	0	3	0	**0**

4

Marks | KU | RE

7. The diagram below shows Isla McGregor's electricity bill.

ScoPower Electricity				
Ms I McGregor 8 Birch Grove Pineford			Account No: 050621743X	
Statement Date: 20 April 2005		From: 21 Feb 2005	To: 18 Apr 2005	
Present reading	**Previous reading**	**Details of charges**		£
006890	006487	**Box A** [ ] units at 7·567p per unit		[·]
		Standing Charge		9·21
			Sub Total	[·]
			VAT @ 5%	[·]
			Total Charge	[·]

(a) Calculate the number of units used.

Write your answer in **Box A**.

1

(b) Complete the electricity bill by filling in the shaded boxes.

3

[Turn over

8. eSunTours is a holiday company.

Last year's percentage income from Skiing, Summer Tours, Winter Sun and Flights is shown in the pie chart below.

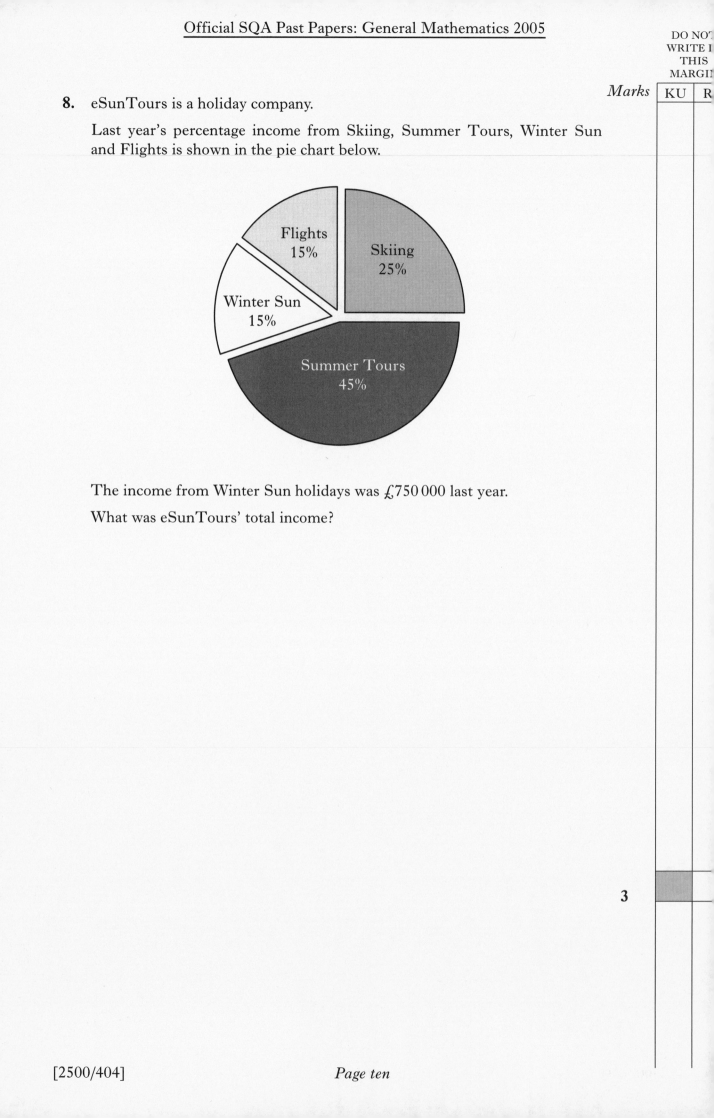

The income from Winter Sun holidays was £750 000 last year.

What was eSunTours' total income?

3

9. Serge drives from his home in Paris to Madrid, a journey of 1280 kilometres.

His car has a 60 litre petrol tank and travels 13 kilometres per litre.

Serge starts his journey with a full tank of petrol.

What is the least number of times he has to stop to refuel?

Give a reason for your answer.

Marks

3

[Turn over

Marks   KU   R

10. (a) The edge of a stock cube measures 1·5 centimetres.

Calculate the volume of the stock cube.

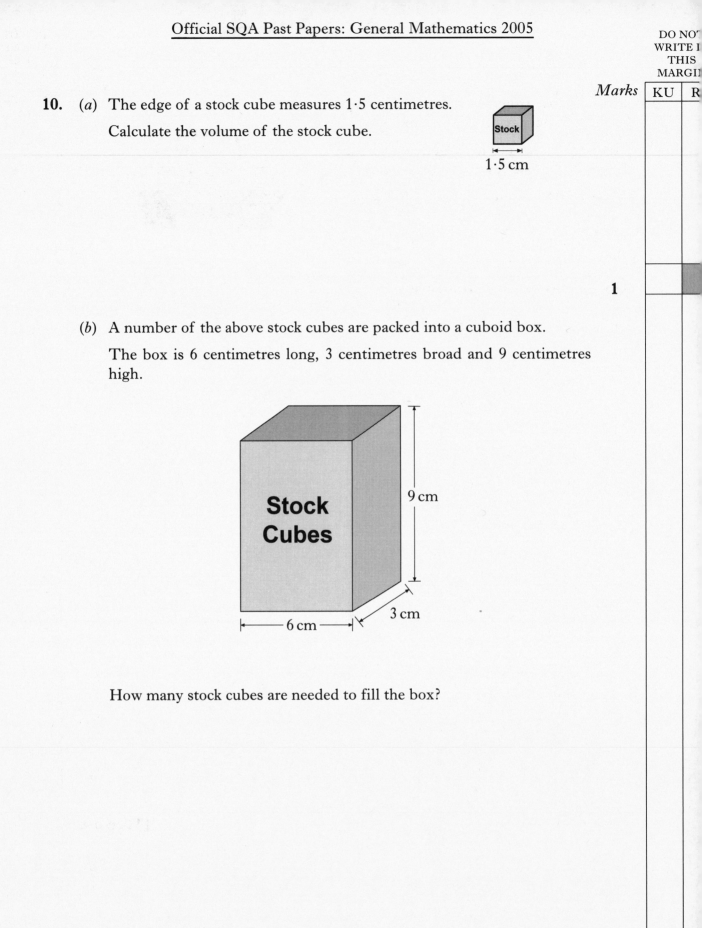

1·5 cm

1

(b) A number of the above stock cubes are packed into a cuboid box.

The box is 6 centimetres long, 3 centimetres broad and 9 centimetres high.

How many stock cubes are needed to fill the box?

3

11.

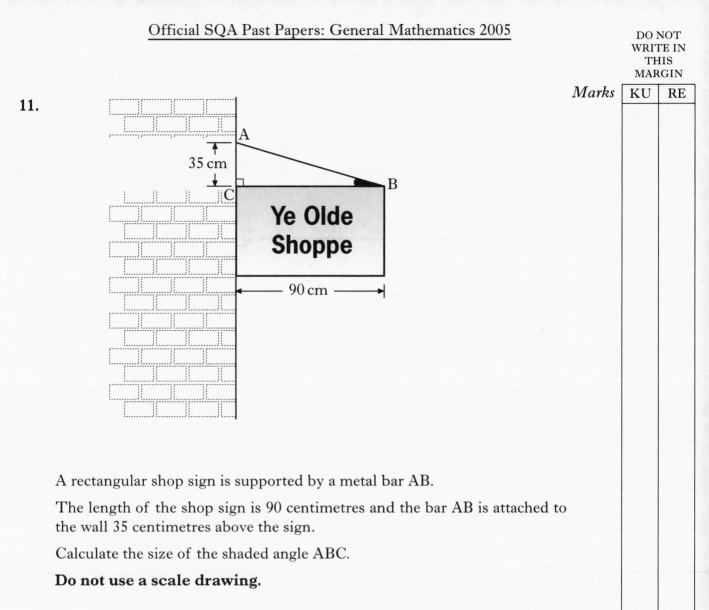

A rectangular shop sign is supported by a metal bar AB.

The length of the shop sign is 90 centimetres and the bar AB is attached to the wall 35 centimetres above the sign.

Calculate the size of the shaded angle ABC.

Do not use a scale drawing.

3

[**Turn over for Question 12 on *Page fourteen***

DO NOT WRITE IN THIS MARGIN

Marks KU RE

12. The diagram below shows the fan belt from a machine.

The fan belt passes around 2 wheels whose centres are 30 centimetres apart.

Each wheel is 8 centimetres in diameter.

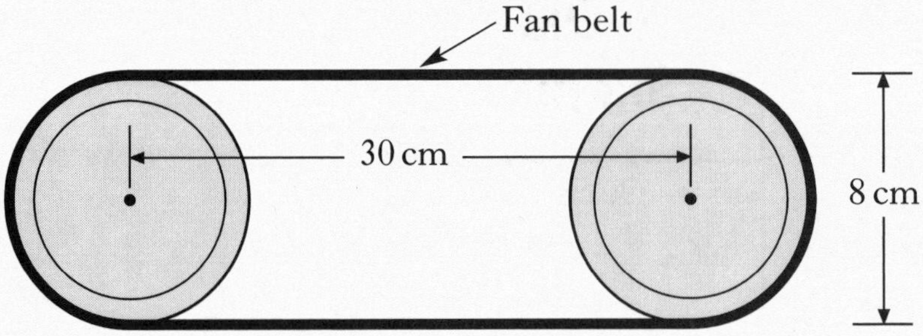

Calculate the total length of the fan belt. **4**

[*END OF QUESTION PAPER*]

ADDITIONAL SPACE FOR ANSWERS

ADDITIONAL SPACE FOR ANSWERS

[BLANK PAGE]

[BLANK PAGE]

[BLANK PAGE]

[BLANK PAGE]